A R R O Y O   C E N T E R

T0168797

MEETING
PEACE OPERATIONS'
REQUIREMENTS
WHILE MAINTAINING
MTW READINESS

P R E P A R E D   F O R   T H E

U N I T E D   S T A T E S   A R M Y

Jennifer Morrison Taw | David Persselin | Maren Leed

RAND

Approved for public release; distribution unlimited

This monograph highlights the final results of a project examining the effects of preparation and deployment to peace operations on U.S. Army combat readiness. It was prepared as part of the "Requirements of Peace Operations" project being conducted for the Army Office of the Deputy Chief of Staff for Operations. The results should be of interest to anyone concerned about readiness issues or interested in peace operations requirements.

The research was conducted in the Strategy and Doctrine Program of RAND's Arroyo Center, a federally funded research and development center sponsored by the United States Army.

CONTENTS

Preface   iii

Tables   vii

Summary   ix

Acknowledgments   xiii

Abbreviations   xv

Chapter One
    INTRODUCTION   1
    Background   1
    Project Objectives   7
    Methodology   7
    Organization   9

Chapter Two
    FORCE STRUCTURE   11
    Introduction   11
    Emphasis on Combat Organization Is Often Inappropriate
        for POs   11
    Deploying Partial Units to POs Stresses MTW Readiness . .   14
    Cross-Leveling for POs Stresses MTW Readiness   16
    High OPTEMPO for AC CS/CSS Units Reduces MTW
        Readiness   18
    Reliance on Volunteers Limits PO Performance   22
    Reserve Activation for POs Can Also Be Problematic   23
    Conclusions and Recommendations   25

Chapter Three
    TRAINING   31
    Introduction   31
    PO Deployments Provide Both Training Benefits and
        Costs   32
    PO Deployments Interrupt Training Cycles   35
    Unmet PO Training Requirements Can Undermine PO
        Efforts   36
    Unmet PO Training Requirements Result in Lower
        Morale   38
    Army Relies on Just-in-Time PO Training   40
    Partial Unit Deployments Impair PO Predeployment
        Training, Interrupt Combat Training Cycles   41
    Cross-Leveling Interrupts Combat Training Cycles,
        Impairs PO Predeployment Training   42
    Conclusions and Recommendations   42

Chapter Four
    EQUIPMENT ACQUISITION, SUPPLY, AND
    MAINTENANCE   47
    PO Deployments Can Erode MTW Equipment Readiness .   47
    Cross-Leveling Equipment and Partial Unit Deployment
        Can Weaken Stay-Behinds' Readiness   50
    Make-Do Equipping Can Harm Both PO Performance and
        MTW Readiness   50
    POs' Equipment Requirements Differ From MTWs'   52
    Conclusions and Recommendations   55

Chapter Five
    CONCLUSIONS   57
    Establishment of a Specialized PO Force Would Be Ill
        Advised   59
    Declining Resources Continue to Pose Challenges   61

Appendix: PROJECT DATABASE   63

Bibliography   65

TABLES

1.1. Sample Deployments of Army Forces to MOOTW   4
1.2. Declining Army Force Structure, 1989 to 1997   5
1.3. Case Study Summary   8
2.1. Army Reserve Component Contribution of Selected
      Units to the Total Army   20

SUMMARY

This project examined the dilemma currently facing the U.S. Army: to prepare for and fight the nation's wars (its primary mission) while also preparing for and conducting peace operations (POs). If it prepares more intensively for POs (those deployments it is actually undertaking), those efforts are unlikely to translate into greater readiness for its primary mission and, in many cases, will result in a degradation of conventional combat readiness. The analysis is based on a series of case studies, an extensive literature review, and interviews with U.S. Army personnel representing combat service and combat service support (CS/CSS), combat arms, and special operations forces (SOF) units; many of those we interviewed were veterans of one or more of the operations in Somalia, Haiti, Bosnia, and Macedonia.

Focusing on the Army's Title X functions to organize, train, and equip, we assessed how the requirement for major theater war (MTW)[1] readiness constrains PO preparation and deployment and, conversely, how PO deployments affect MTW readiness.[2] We also

---

[1]The term "major regional contingency" (MRC) has become outdated since this report was written. MRC was originally used to indicate a large-scale conventional conflict short of a global war. Operation Desert Storm, for example, was an MRC. The new term of art, introduced in the 1997 Quadrennial Defense Review, is "major theater war" (MTW), and we use it here.

[2]We explicitly discuss preparation for and deployment to peace operations, but we use the term "deployment" in reference to the entire deployment cycle, including employment and recovery. Indeed, as will be discussed in greater detail later in the report, recovery is a critical aspect of this question; how long it takes and what it

examined how the Army's current focus on MTWs determines how it conducts POs. We found that despite the increased numbers of POs, the Army does little routine preparation for peace operations.

KEY FINDINGS

This analysis produced five key findings. First, deployment to POs has reduced MTW readiness for certain frequently deployed and/or low-density unit types (military police, civil affairs, psychological operations, and certain transportation units, for example).[3]

Second, since little preparation for POs is undertaken, it is deployments to—not preparation for—POs that have diminished some units' MTW readiness.

Third, PO deployments' effects are dispersed beyond units that have deployed, are preparing to deploy, or have returned from deployment: cross-leveling and interruptions of collective training affect stay-behind units as well.[4]

Fourth, the situation could get worse. Thus far, the Army's role in peace operations has been relatively limited and soldiers have been able to accomplish their PO missions. But because of the Army's pool of equipment and skilled labor—and a concomitant lack of organized civilian capabilities on the same scale—there has been pressure in each operation for the Army to become further involved in nation-building. Although disparaged as "mission creep" by some,

---

requires to return troops and units to readiness for their primary mission following a peace operations deployment is a subject that remains highly contested.

[3]For a list of heavily deployed MOSs, see U.S. Department of Defense (1996). Also, consider the effects of PO deployments on the 3rd Psychological Operations Battalion (POB), the Army's only active technical PSYOP battalion. In May 1996, this battalion was authorized 326 personnel but had only 200 on hand. Of those, 49 were in Bosnia, 6 were in Haiti, and another one was participating in Operation GTMO. Yet in a two-MTW scenario, MTW 1 calls for 186 personnel from 3rd POB, and MTW 2 for another 186. Clearly, at this battalion's very constrained manpower levels, the deployments to POs compromised even the ability to meet MTW 1's projected requirements. "3d POB (A) PSYOP PERSONNEL CAPABILITY," unpublished document, May 1996.

[4]That peace operations deployments have a ripple effect is undeniable (indeed, any deployment will have a ripple effect, as long as units are maintained at less than full strength); however, it is debatable how important that effect really is on the Army's overall ability to wage war effectively.

others consider greater Army involvement in civic action, humanitarian assistance, and civil-military relations to be key both to force protection and to creating the required end state. If the Army's role in peace operations is expanded to include more nation-building, some adjustments in force structure, equipment, training, and doctrine will have to be made. Personnel from transportation, medical, military police, vertical engineering, legal, special forces, civil affairs, and psychological operations units—among others—will have to be made available in sufficient quantity, with adequate command elements, appropriate training, and specialized equipment. In a zero-sum resource environment, such preparation for POs would further diminish MTW readiness.

An oft-cited alternative would be to create a special PO force within the Army, but our research indicates that such an option, given the scope and depth of PO requirements, is unrealistic and should be rejected.

Finally, there are some win-win solutions that can both improve PO performance and mitigate some of the adverse effects of PO deployments on MTW readiness. Such steps include (1) building greater flexibility into the force structure, (2) relying more on other agents (joint, interagency, coalition, and private), (3) maintaining MTW skills by training during PO deployments, (4) emphasizing leader education, and (5) deploying a single set of equipment for duration of the PO.

ACKNOWLEDGMENTS

The authors would like to thank LTC Wyman Irwin, the 1996 senior RAND Army Fellow, and Lois M. Davis, a RAND consultant with expertise on the medical aspects of military operations other than war, both of whom participated in the project's early stages of research and analysis. The authors are also grateful to the many individuals who took time from their busy schedules to share their opinions and experiences. In particular, the authors would like to thank the participating personnel from Forces Command, Atlantic Command, the 4th Psychological Operations Group, the 3rd Special Forces Group, the 96th Civil Affairs Battalion, the 519th Military Intelligence Battalion and the 525th Military Intelligence Brigade, the 16th Military Police Brigade, the 20th Engineer Brigade, and the Engineer School, as well as COL Tom Molino of DAMO-SSP, Stuart Drury of DAMO-FDF, and LTC Larry Moores of DAMO-SSW. Additionally, a number of RAND colleagues offered substantial and valuable assistance and advice during the course of this project. Key among them were Tom McNaugher, Michael Polich, Ron Sortor, Tom Szayna, Jim Quinlivan, John Dumond, and Rick Eden. Any errors of fact or analysis are, of course, the authors' own.

ACOM	Atlantic Command
AC	Active Component
CA	Civil Affairs
CALL	Center for Army Lessons Learned
CASCOM	Combined Arms Support Command
CFP	Contingency Force Pool
CMTC	Combat Maneuver Training Center
COIN	Counterinsurgency
CONUS	Continental United States
COSCOM	Combat Support Command
CS/CSS	Combat Support/Combat Service Support
CSB	Corps Support Battalion
CT	Combatting Terrorism
EAC	Echelons Above Corps
EAD	Echelons Above Division
FM	Field Manual
FORSCOM	Forces Command
FSP	Force Support Package

GAO	General Accounting Office
GTMO	Guantanamo
IFOR	Implementation Force (Bosnia)
JLOTS	Joint Logistics Over-The-Shore
JRTC	Joint Readiness Training Center
LOGCAP	Logistics Civil Augmentation Program
METL	Mission-essential Task List
MFO	Multinational Force and Observers
MI	Military Intelligence
MOOTW	Military Operations Other Than War
MOUT	Military Operations on Urban Terrain
MOS	Military Occupational Specialty
MP	Military Police
MRC	Major Regional Contingency
MTW	Major Theater War
MTT	Mobile Training Team
NATO	North Atlantic Treaty Organization
NEO	Noncombatant Evacuation Operation
NTC	National Training Center
OCONUS	Outside Continental United States
OPTEMPO	Operating Tempo
OUD	Operation Uphold Democracy
PE	Peace Enforcement
PEO	Peace Enforcement Operation
PK	Peacekeeping
PO	Peace Operation
PSRC	Presidential Selective Reserve Call-up

PSYOP   Psychological Operations

RC   Reserve Component

SF   Special Forces

SOF   Special Operations Forces

TAA03   Total Army Analysis 2003

TAV   Total Asset Visibility

TOE   Table of Organization and Equipment

TPFDD   Time-phased Force and Deployment Data

TPFDL   Time-phased Force and Deployment List

UAV   Unmanned Aerial Vehicle

UIC   Unit Identification Code

UN   United Nations

UNPROFOR   United Nations Protection Force

USAREUR   U.S. Army Europe

Chapter One

INTRODUCTION

> We need to be very careful that this [peace operations]
> does not become our way of life; that we remember that we
> are first and foremost to fight our nation's wars.
>
> —General John Shalikashvili[1]

BACKGROUND

This project examined the dual challenge facing the U.S. Army of
preparing and deploying frequently for peace operations (POs) while
maintaining readiness for fighting and winning the nation's wars, its
primary mission. If it prepares more intensively for POs (those deployments it is actually undertaking), its efforts are unlikely to translate into greater readiness for its primary mission and, in some cases,
will result in a degradation of MTW readiness.

The Nature of Peace Operations

In Army Field Manual (FM) 100-23, *Peace Operations*, peace operations are defined as activities that "create and sustain the conditions
necessary for peace to flourish." Included within this rubric are support to diplomacy, peacekeeping (PK), and peace enforcement (PE).
This project focused specifically on the latter two kinds of peace operations, which are defined in FM 100-23 as follows:

---

[1]Komarow (1996).

> Peacekeeping . . . involves military or paramilitary operations that are undertaken with the consent of all major belligerent parties. These operations are designed to monitor and facilitate implementation of an existing truce agreement and support diplomatic efforts to reach a long-term political settlement.

> [Peace enforcement] is the application of military force or the threat of its use, normally pursuant to international authorization, to compel compliance with generally accepted resolutions or sanctions. The purpose of PE is to maintain or restore peace and support diplomatic efforts to reach a long-term political settlement.[2]

The Army treats peace operations doctrinally as a subset of the broader category of military operations other than war (MOOTW),[3] which includes such disparate activities as noncombatant evacuation operations (NEOs), combatting terrorism (CT), counterinsurgency (COIN), arms control, and attacks and raids. We deliberately did not include either support to diplomacy or the full range of MOOTW in this study: the effects of disaster relief efforts in Rwanda and NEOs in Liberia, for example, were excluded. Peacekeeping and peace enforcement are differentiated from other MOOTW by the Army forces used and the flexibility of the missions and durations. For example, other MOOTW, such as counterdrug, noncombatant evacuation, and counterinsurgency operations, usually employ specialized forces or conventional forces trained specifically for such missions. Peace operations, in contrast, have employed large numbers of conventional infantry and support forces, in addition to the more specialized units. Because they draw on many of the same forces that MTWs require, POs, unlike less conventional MOOTW, are arguably most likely to stress Army resources and undermine readiness for a major war.

---

[2]Department of the Army (1994), pp. iv, 4, 6.

[3]Department of the Army (1993a), pp. 13-0 to 13-8. For a discussion of the term "military operations other than war" and its genesis, see Story and Gottlieb (1995). The Army has already begun to reject MOOTW and replace it in draft doctrine with "stability and support operations" (SASO). The intention is to move away from a separate category of operations, instead defining operations by tasks as offensive, defensive, support, or stability operations.

Indeed, while Army participation in the more traditional MOOTW continues (in 1995 alone, the Army supported 553 continental U.S. (CONUS) and outside CONUS (OCONUS) counterdrug operations,[4] for example, each of which involved very limited numbers of specialized Army troops), peace operations such as Able Sentry in Macedonia, Restore and Continue Hope in Somalia, Uphold Democracy in Haiti, and Provide Comfort in Turkey have involved large numbers of soldiers over prolonged periods (see Table 1.1).

The U.S. Army Concepts and Analysis Agency's 1991 Force Employment Study clearly illustrates the magnitude of effort that peace operations can require. In a study of 49 operations undertaken between 1975 and 1991, the single peacekeeping operation of the period (the multinational force and observers (MFO) in the Sinai) accounted for 30 percent of the total documented man-day allocations and almost 39 percent of total allocated man-days for OCONUS operations.[5]

Army Organizes, Trains, and Equips for Combat

The post–Cold War drawdown of budgets, manpower, and infrastructure makes the conduct of POs while maintaining MTW readiness a pressing issue. Although the Army has faced similar operational challenges since the end of World War II, until recently it has had sufficient resources to conduct contingencies short of war as "lesser-included cases." It assumed that preparation for war provides the capabilities for any other kind of operation that might arise. In the wake of the Cold War, by contrast, the Army has become smaller, and it has transitioned from a force that based significant portions of its strength overseas to one that will conduct power-projection operations from CONUS and a few other locations (see Table 1.2).

---

[4]U.S. Department of Defense (1996).

[5]U.S. Army Concepts Analysis Agency (1991), pp. vi, 1–3. This number must be kept in perspective: it represents the long-term engagement of a small fraction of the total force at a time when the force was bigger and there were far fewer operational and training deployments than there are today. Nonetheless, it is not insignificant that a single operation could account for such a high percentage of allocated man-days in a period that included force deployments for two wars (in Panama and Grenada) and myriad operations other than war.

Table 1.1

Sample Deployments of Army Forces to MOOTW

Operation	Peak Number of Army Troops Deployed	Sample Dates
Restore Hope[a]	9,608	September 92–April 94
Uphold Democracy[b]	11,563	August 94–February 95
UN Mission in Haiti (UNMIH)	3,530	April 95–April 96
Joint Endeavor[c]	24,143	December 95–August 96

NOTE: These data represent peak numbers of troops deployed to these operations between the specified dates as reported by the Joint Staff J1 Manpower and Personnel Directorate. The actual peak for the operation may have occurred before or after the listed dates, but the numbers shown nonetheless sufficiently demonstrate the large number of Army forces deployed for these more conventional MOOTW.

[a]Based on TPFDD for Operation Restore Hope.

[b]Based on TPFDD for Operation Uphold Democracy.

[c]Operation Joint Endeavor was the NATO operation to implement the Dayton Peace Agreement in Bosnia.

At the same time, the amount of time soldiers spend deployed (operational tempo, or OPTEMPO) has actually increased, driven by, among other things, deployments for peace operations, humanitarian assistance, disaster relief, and support to counterdrug operations, as well as more joint and combined exercises.[6] Indeed, between the late 1980s and early 1995, the portion of the Army force deployed for operations or exercises increased from an average of 5 percent to approximately 8.5 percent.[7] This tension between growing operational requirements and decreasing resources limits the Army's ability to respond to unanticipated contingencies.

---

[6]Between 1989 and 1996 the number of Army soldiers on active duty has decreased from 781,000 to 495,000. During that time, there have been 25 "major" deployments. Yet from 1950 to 1989 there were only 10 "major" deployments (including the Korean and Vietnam wars). Of course, the 10 major deployments and the buildup of forces to the 781,000-person high took place in the context of the Cold War and the threat of global conflict; since 1989, that threat has been substantially eased, and the Army is preparing more for regional than for global warfare. Fisher (1996), p. 1.

[7]U.S. General Accounting Office (1996a), pp. 4, 9. The report cites the drawdown in personnel, the reduction in overseas presence, and the large increases in joint activities since the end of the Gulf War as key factors affecting deployment tempo.

Table 1.2

Declining Army Force Structure, 1989 to 1997

Year	Active Divisions	Active Endstrength	Percent of Force Based OCONUS
1989	18	770,000	42.3%
1997	10	495,000	24.8%

SOURCE: West and Reimer (1997).

The Army's flexibility to undertake "lesser-included cases" is declining just as deployments to relatively conventional, frequently prolonged, and financially costly[8] peace operations are on the rise. Although the Army participated in only two operations that can be cast even in the broadest terms as peace operations during the 44 years of the Cold War (in the Dominican Republic and Egypt),[9] there have been six such operations (Iraq, Somalia, Haiti, Macedonia, and Bosnia, in addition to MFO Sinai) in the past seven years alone.

Nonetheless, since the end of the Cold War the Army has continued to treat peace operations, for the most part, as "lesser-included cases." It has made some adjustments to doctrine and training to meet peace operations' needs, but, in response to the national security and military strategies, its primary focus remains preparation for MTWs. This focus has driven even the most recent Army planning and organization efforts. For example, *Total Army Analysis 2003* (TAA03) identified key CS/CSS capabilities that are being strained by current operations.[10] Rather than trading off combat troops in order

---

[8]The General Accounting Office reported in March 1996 on the fiscal year 1995 incremental costs to DoD of the following peace operations: Somalia, $49 million; Bosnia (to that point), $347 million; Haiti (to that point), $569 million; and Rwanda, $36 million. U.S. General Accounting Office (1996b), p. 8.

[9]The three peacekeeping efforts in Lebanon in which the United States participated during this period—in 1958 and again in 1982–1983—involved Marines, not Army forces.

[10]For a table showing the TAA03 shortages in authorized positions for a variety of branches (mostly quartermaster, transportation, air defense, signal, and engineer), see U.S. General Accounting Office (1997a), p. 31.

to augment such capabilities, however, CS/CSS capabilities were traded off against each other, because combat troops are fenced.[11]

Competition for Resources

If PO and MTW resource requirements were identical, the question would simply be how PO deployments depleted the resources available for MTWs. However, in addition to drawing from MTW-designated resources, POs have unique force structure, training, and equipment requirements. This further complicates the Army's efforts to set priorities. Expenditures and adjustments made specifically to meet PO needs will usually require trading off MTW capabilities.

Some may question how real—or important—the tradeoff is in light of the apparently benign post–Cold War security environment. Interviews with senior Army personnel indicate that they consider the question in different terms: Where can the risks be taken? If the Army prepares for war, and therefore conducts POs suboptimally, it assumes one kind of risk, usually political. If it prepares for POs, and then cannot fight a war, it assumes a risk at a higher order of magnitude. Interviewees also considered the Army's capacity for ad hoc response to one situation versus the other: combat capabilities given up now might take years, even decades, to rebuild, whereas ad hoc responses to PO requirements have proved sufficiently effective, if not ideal.

Given this, it would be helpful if the Army could turn to its accumulated experience in, and preparation for, other MOOTW to reduce the strain on resources imposed by greater involvement in POs, rather than relying on resources earmarked for MTWs. Unfortunately, this is not the case. Indeed, insofar as other MOOTW also require special operations forces and specialized CS/CSS capabilities, deployments to them require many of the same key MTW capabilities strained by POs. This is particularly true for disaster relief, humanitarian assistance, and nation assistance, as well as for COIN and CT.

_____

[11]Authors' interview with Stuart Drury, DAMO-FDF.

Additionally, preparation for yet other MOOTW can result in training and equipment expenditures that benefit neither MTWs or POs. For example, the 1st PSYOP Battalion (one of only five active PSYOP battalions) conducts mostly counterdrug training. It has not practiced for its intended role as the Psychological Operations Task Force in the event of two near-simultaneous MTWs. Nor did 1st PSYOP Battalion personnel receive special training before deploying to Somalia or Operation Joint Endeavor, which would have refocused their skills for the PO mission. Involvement in other MOOTW thus resulted in lost opportunity costs for either POs or MTWs.[12]

PROJECT OBJECTIVES

In response to concerns about how the Army can meet the demands of both POs and MTWs, we sought first to identify PO-specific requirements, to assess whether they had been met in past operations, and to determine how crucial they might be to satisfying operational objectives.

We next examined how preparation for and deployment to peace operations stressed or enhanced readiness for an MTW, framing the question specifically in the context of the Army's Title X responsibilities.

We then compared how MTW readiness would be affected by preparing for and deploying to POs as "lesser-included cases" versus how it would be affected by meeting PO-specific requirements through adjustments to force structure, training, and equipment.

The project's final objective was to identify steps the Army could take to maximize soldiers' effectiveness (as individuals and as units) in both POs and MTWs.

METHODOLOGY

Our research effort involved an extensive literature review, case studies, and interviews with Army personnel. The literature review in-

---

[12]Authors' interview with 1st PSYOP Battalion personnel, Fort Bragg, North Carolina, May 16, 1996.

cluded not only academic and professional articles and books, but Army doctrine and reports and studies from such agencies as the U.S. Army Concepts Analysis Agency, the Center for Army Lessons Learned, the Combined Arms Center, the TRADOC Analysis Center, the General Accounting Office, the Congressional Budget Office, and Science Applications International Corporation.[13]

The case studies we chose—U.S. operations in Bosnia (U.S. involvement in the UN Protection Force—UNPROFOR—and the NATO Implementation Force—IFOR), Haiti (leading up to and through Operation Uphold Democracy), Macedonia (Able Sentry), and Somalia (Operations Restore and Continue Hope)—offered the opportunity to examine both peacekeeping and peace enforcement (see Table 1.3). In addition, by looking at U.S. operations in each location over time, we were able to study how POs change and evolve.

Finally, we conducted interviews with Army personnel from CS/CSS, combined arms, and SOF units, many of whom were veterans of one or more peace operations. The interviews took place with personnel at Fort Bragg, Fort Leonard Wood, Combined Arms Support Command (CASCOM), Atlantic Command (ACOM), Forces Command (FORSCOM), and U.S. Army Europe (USAREUR), from combat arms,

Table 1.3

Case Study Summary

Location	Dates	Major Army Units	Operations
Somalia	1992–1994	10th Mountain Division	Restore Hope Continue Hope
Macedonia	1993–present	Infantry Battalion TF Rotation	Able Sentry
Haiti	1994–1996	10th Mountain Division 25th Infantry Division	Uphold Democracy UNMIH
Bosnia	1995–present	1st Armored Division 1st Infantry Division	Joint Endeavor Joint Guard

---

[13]For a useful annotated bibliography of relevant studies and literature, see Department of the Army, TRADOC Analysis Center (1996).

engineer, signal, military intelligence (MI), special forces (SF), psychological operations (PSYOP), civil affairs (CA), quartermaster (QM), military police (MP), and transportation units, among others.

To store and organize our extensive notes, we built a qualitative database. Each entry is classified by unit type, mission, source, and general and specific topics. This database, containing nearly 1,000 entries, allowed us to look across types of units and missions—as well as individual data sources—to identify and evaluate issues common to a variety of circumstances as well as those that might be unique to a particular mission or unit type (or even to an individual).[14]

ORGANIZATION

This monograph is divided into five chapters. Chapters Two through Four discuss how the nation's—and the Army's—emphasis on combat readiness affects both MTW readiness and PO performance in terms of force structure, training, and equipment. The final chapter discusses how the Army has responded to POs to date, and how it might better prepare for the challenges that may arise in the future.

---

[14]Whenever possible, we sought data that might provide additional quantitative insights. However, until recently the Army did not collect much of the information relevant to our analysis; while this situation may be improving and more complete data in the future will hopefully shed further light on the complex effects of POs, we sought to use the available qualitative information in a rigorous and thorough manner. We believe this approach has led to some robust conclusions that would benefit from additional quantitative work, but that are of value in their own right as well.

Chapter Two

FORCE STRUCTURE

INTRODUCTION

Army forces are currently sized and organized to ensure their ability to meet the requirements of two nearly simultaneous MTWs. This force structure is not, however, ideally suited to POs. This chapter examines how structuring for MTWs while deploying for POs ultimately affects both MTW readiness and PO capabilities.

EMPHASIS ON COMBAT ORGANIZATION IS OFTEN INAPPROPRIATE FOR POs

The Army's organization into corps and divisions is designed to maximize combat performance in an MTW. For example, the functional building blocks of a division and corps support system (transportation, supply, and maintenance) are scaled so that, when combined, they have the capacity to support single or multiple division-level campaigns. However, the expectation of divisional or corps-level deployments has not been met in most POs. Thus, tailoring a force for POs can be particularly challenging.[1]

The small scale of POs relative to MTWs, and/or the imposition of force caps (such as the stated 20,000 troop limit in Bosnia[2]), have re-

---

[1]Some analysts might note that force tailoring for smaller combat operations (such as Just Cause in Panama) is equally problematic, given the distribution of capabilities at echelons above division or above corps.

[2]This limit was perceived as unrealistic, given the anticipated operational requirements. Additional U.S. forces were therefore placed in neighboring countries, such as Hungary.

quired or allowed only a one- or two-brigade task force in most operations, and only selected elements of support units designed for division- or corps-level operations have been deployed. Despite the smaller scale of most POs, however, it is still appropriate for PO planning to be done at corps level because of the requirements to plan for and coordinate among higher-echelon forces, other services and government agencies, coalition partners, contractors, and nongovernmental organizations (NGOs).[3]

POs emphasize policing, building, transporting, and facilitating rather than combat arms functions, and therefore require CS/CSS and SOF forces in higher ratios to combat arms personnel than an MTW would require.[4] Conventional Joint Task Force staffs—historically formed around combat arms units—are not well prepared by either training or experience to provide planning or command and control for what is essentially a CS/CSS/SOF mission with a force protection and observation component, as was the case in Haiti.

Under such circumstances, one proposed solution is to have professional logisticians, engineers, civil affairs personnel, or police run the operation in lieu of infantry officers.[5] Some special forces personnel

---

[3]For a discussion of the reasons for using a corps, rather than division, staff as the base for a JTF staff, see U.S. Army Center for Army Lessons Learned (1993–1994), p. I-2-6. The division task force headquarters may be augmented, however, to cope with the unusual task organization and coalition coordination requirements of many POs. Each of the operations examined for this study, for example, was top-heavy. One observer defended the similarly organized MFO Sinai, claiming that "it is a justifiably large organization, and it is unfair to assert as some do, that the current force headquarters wield a corps-size staff to command less than a brigade deployed in the field, for no brigade headquarters is equipped to cope with the peculiar MFO problems of international liaison." MacKinlay (1983), p. 58.

[4]For example, Major General Kinzer, the UN Force Commander and commander of the U.S. forces in Haiti, noted that in this mission there was high demand for linguists, military police, engineers, medics, and logistical support. Niblack (1995), p. 33. Likewise, a study of the 1965 operation in the Dominican Republic found that a disproportionate share of CS/CSS would be required, specifically, "[s]ignal, military police, medical, and logistical units." *United States Stability Operations in the Dominican Republic 28 April 1965–30 May 1965*, Part I, Volume IV, Chapter 17: "Doctrine and Force Organization, Headquarters, U.S. Forces Dominican Republic, August 31, 1965," p. J-5, cited in Reese (1987), p. 26.

[5]This was a major topic of discussion at the United Nations Mission in Haiti (UNMIH) U.S. Army After Action Review held at the U.S. Army War College in May 1996. In that discussion, it became apparent that although some of the command and control (C2) issues arose from difficulties coordinating between UN and U.S. organizations,

argue, for example, that they have all the skills and experience to run peace operations, although they question whether they would want such a mission.[6] Likewise, in the view of one logistics officer,

> Logistics organizations in the U.S. military have the resident expertise to accomplish peacekeeping operations more efficiently than combat troops. These organizations are flexible and contain modular units that can be reconfigured easily to support any mission. Embedded in these organizations are medical, distribution, supply, and communications assets, all of which are essential in the execution of peace operations . . . . Senior logisticians have the experience, knowledge, and leadership skills required to supervise peace operations. They understand the capabilities and constraints of military equipment and can successfully provide logistics support to large organizations in austere environments.[7]

The dilemma lies in peace operations' unpredictability and changeability over time. Whereas the force protection and security aspects of an operation may predominate at a given time, under which circumstances a combat arms officer will most likely be better suited to plan and employ the force, at another time the nation-building re-

---

questions also arose about how well a combat arms–based structure could integrate and coordinate logistical, signal, MP, engineer, CA, SF, and PSYOP personnel in POs. C2 issues in POs were also raised at the Combat Maneuver Training Center, where rotations in April and May 1993 demonstrated that task forces did not sufficiently integrate or employ military police or special staff elements such as CA personnel, interpreters, and liaison officers. Seventh Army Training Command (1993), pp. 12, 30, 45. A Joint Universal Lessons Learned (JULLS) report cited a similar problem in Somalia, where the civil affairs teams did not arrive until D+20. The report called for CA teams to join their supported units in CONUS, conduct predeployment training with them, and begin staff coordination. JULLS No. 10704-59654 (00097). An alternative to having such specialized forces *running* operations is to improve their representation and influence on JTF staffs. Staff organization for these operations, as for typical combat operations, has tended to accord more influence to the operations officers, then to the logistics officers, and, in the distance, to the political-military officers. Specialized branches have sometimes had very junior representatives on the JTF staff, or have been buried deep in the staff structure, several layers beneath the staff officers.

[6]Authors' interview with 3rd Special Forces Group personnel, May 16, 1996, Fort Bragg, North Carolina.

[7]Pilgrim (1996), pp. 38–39. Logistics organizations, such as a COSCOM or a DISCOM, are not currently authorized the personnel and equipment to provide command and control for a fully synchronized PO, but these shortcomings could be resolved with augmentation from either the active or reserve components.

quirements may take precedence, challenging the combat arms officer's knowledge of, and familiarity with, the support and special operations forces' capabilities. Thus, one issue is whether the same command or staff structure should—or can—be used throughout the operation as it permutates.

Another issue is whether a division task force—which will typically rely primarily on its organic CS and CSS capabilities—has the skills or equipment required to perform the tasks in a PO. For example, divisional engineers have the equipment and training to breach minefields and obstacles and dig fighting positions—but POs call for horizontal and vertical construction capabilities, which reside in EAD units.[8] Similarly, divisional intelligence units specialize in counterintelligence and interrogation, rather than the human intelligence (HUMINT) capability that is more useful in POs but is found primarily at higher echelons. So the Army's predisposition to structure forces for POs as if they were combat operations has sometimes led to suboptimal mission tailoring and reliance on inappropriate assets.

DEPLOYING PARTIAL UNITS TO POs STRESSES MTW READINESS

The task forces created for POs are often characterized by the deployment of partial, rather than whole, units. These types of deployments tend to weaken both elements of the severed unit. To prepare for the mission, the deployed element typically takes a disproportionate amount of the unit's key leaders, low-density MOS personnel, and serviceable equipment and, to the extent that the unit's role in the PO mission requires planning and coordination activities, may also draw critical command and control assets away. Not only are the most capable personnel often deployed, but frequently those who are less deployable (because of preexisting medical conditions, for example, or family difficulties) stay in the rear detachment, further undermining its readiness.[9] These practices

---

[8]Authors' interview, Engineer School, Fort Leonard Wood, Missouri, May 16, 1996. This was also a recommendation of the Seventh Army Training Command, after observing a series of CMTC rotations in April and May 1993. Seventh Army Training Command (1993), p. 23.

[9]Authors' interview with CPT Bunch, Fort Bragg, North Carolina, June 26, 1996; Ritchie, Ruck, and Anderson (1994), p. 376.

degrade the readiness of the stay-behind elements by stripping them of the leadership, skills, and equipment they would need to rapidly deploy to an MTW.[10] For example, one signal battalion deployed only one of its companies to Haiti, which resulted in a shortage of 31-series MOS communications specialists, dropping the battalion's readiness status to C-4.[11]

Splitting units can also result in problems for the elements that actually deploy. Just as sending key assets cripples rear detachments, holding them at home station can inhibit the effectiveness of forward elements. In Somalia, for example, force caps imposed specifically on the 10th Mountain Division led its aviation brigade to deploy a lift company without any battalion command and control; instead, a larger force package was deployed from Europe, where force caps were not in effect.[12] But relying on unfamiliar units for support, and particularly for command and control, increases the chance of failures due to miscommunication and lack of shared procedures.

In terms of the MTW mission, the result of partial unit deployments is that some units are not fully capable of rapidly deploying in the event of a major conflict. The assumption that the PO-deployed element can quickly redeploy to an MTW theater is questionable. If stay-behind units are unable to quickly marry up with their deployed elements in the event of an MTW, the commander will be forced to choose between sending a unit that is not fully manned and sending a unit less suited to the mission. Substituting units will cause command and control problems if they lack the operational relationships with the supported unit developed through habitual combined training. Additionally, some military police, transportation, and quartermaster units have installation missions that force commanders to weigh whether those assets are more critical in-theater against the support of other units at home station.[13]

---

[10]"When the approximately 150 of 180 military personnel from the XVIIth Airborne Corps' 507th Combat Support Group Headquarters deployed to Somalia for several months, they left approximately 30 headquarters personnel at Fort Bragg, along with the group's three battalions, without any additional augmentation." U.S. General Accounting Office (1995b), p. 23.

[11]Shelton memo (1995).

[12]Department of the Army (1993b), p. 36.

[13]Authors' interview with CPT Bunch, Fort Bragg, North Carolina, June 26, 1996.

CROSS-LEVELING FOR POs STRESSES MTW READINESS

Because the Army maintains many of its active component (AC) CS/CSS units understrength, once units are selected for deployment, many must be augmented with personnel from like units in order to reach full strength. For its deployment to Haiti in Operation Uphold Democracy, 25 percent of the 118th Military Police Company was drawn from other units to obtain enough soldiers of the required grade and skill level.[14] This kind of cross-leveling takes key assets from other AC units, thus disseminating the adverse impact throughout the force. Moreover, in many cases, the lending units are themselves short the same key assets even before the cross-leveling, since like units often experience the same shortages.

The GAO reported in August 1996 that of the thirty-one Army and five Air Force units they reviewed that participated in the Bosnia operation, "five Army units (14 percent) and one Air Force unit (20 percent) reported readiness reductions," and that "the Army units had deployed elements or key personnel to Bosnia, thus lowering resources available to the parent (reporting) units." Again, these are support units: the five the GAO identified were a civil affairs, a signal, a psychological operations, and two transportation units.

The problem of cross-leveling is widespread.[15] It affects even the contingency force support package (FSP) which, under the current strategy, will quickly deploy in the event of an MTW in support of the early-deploying combat arms units.[16] In other words, the FSP is made up of "those [active and reserve component] CS/CSS units designated to support the range of contingency responses that could

---

[14]Authors' interview with officers of the 19th Military Police Brigade, Fort Bragg, North Carolina, May 17, 1996.

[15]"Cross-leveling has occurred at both the division and corps level. For instance, the 210th Forward Support Battalion, an element of the 10th Mountain Division, took people and equipment from the Division's 46th Forward Support Battalion and the 710th Main Support Battalion before deploying to Somalia. The 710th Main Support Battalion also supported the 46th Forward Support Battalion's deployment, thereby creating a domino effect within the 10th Mountain Division. According to the 710th commander, the battalion deployed with fewer than all its people and equipment." U.S. General Accounting Office (1995b), p. 22.

[16]Clearly, if these capabilities are already deployed to, or recently returned from, a PO, this strategy may be complicated. The implications of the Army's current AC/RC mix for POs and, in turn, for MTWs, are profound, and will be discussed in Chapter Three.

occur in a crisis response."[17]   Yet FSP units in FORSCOM are currently providing individual augmentees to USAREUR in support of
Operation Joint Endeavor, and in many cases these are soldiers who
by their duty position or skills are critical to the combat effectiveness
of their unit. These same soldiers are not realistically available to
deploy with their unit in the event of an MTW. One brigade S-3 reported that he had just returned from 179 days of temporary duty
(TDY) as an augmentee to the 1st Armored Division's engineer
brigade, during which time he was completely unavailable to train
with his assigned unit—yet in the event of an MTW he would have
been expected to redeploy directly from Bosnia to join his unit and
go into combat.[18] Indeed, the engineer brigade in Bosnia was
augmented with over 40 officers on TDY from their assigned units to
give that headquarters sufficient command and control capability to
coordinate the efforts of all the nondivisional engineer assets assigned to it.[19] Other unit types have experienced similar effects—
PSYOP battalions, for example, have been particularly hard hit by
cross-leveling, as they have only one O-6 and few O-5s.[20]

Such cross-leveling not only hurts combat readiness, it blunts PO
performance by creating agglomerated, uncohered units.[21] In Somalia, for example, the 62nd Medical Group and the Logistical Sup-

---

[17]According to the Army's Force Planning Guidance, "The FSP includes those doc-
trinal forces required to deploy and support 5 1/3 CONUS divisions, EAD/EAC for one
Corps, and support elements to open one theater. Included are those essential forces
to support forcible entry operations and CONUS Base units required to support mobilization and deployment." The FSP has replaced the contingency force pool (CFP), but
the concept is the same.

[18]Authors' interview with MAJ Nicholson, Fort Bragg, North Carolina, May 17, 1996.

[19]Authors' interview with LTC(P) Robert L. Davis, Fort Leonard Wood, Missouri, May
16, 1996.

[20]Similar cross-leveling has taken place within the National Guard as well. Such
"tailoring" has led to the mobilization of partial Guard units and even of individual
soldiers. Haskell (1996).

[21]There is very little data currently available on cross-leveling. Recognizing this as a
problem, however, the Army Personnel Command (PERSCOM) has recently developed
the "Operation Deployment Non-Deployable Report Requirement," which has
directed that "non-deployable tracking and reporting include all operational deployments." Units deploying on operational deployments will have to submit, among
other things, the "Number, rank, and MOS of soldiers cross-leveled from outside
deploying unit to meet deployment manning guidance. Number of replacements
from within the installation [sic]."

port Command were each made up of personnel from all over CONUS and Germany, most of whom did not know each other and were unfamiliar with the command structure.[22] The military police component in Somalia also reflected a high level of cross-leveling: although "the TPFDD contained only ten MP companies . . . [t]he personnel deployment data showed [that] MPs . . . deployed from 62 different UICs."[23]

HIGH OPTEMPO FOR AC CS/CSS UNITS REDUCES MTW READINESS

The Army's combat orientation also dictates the current mix of forces in the active and reserve components, which, though well suited for combat operations, is less ideal for support-intensive POs. The MTW requirement for rapid access to combat power has translated, given limited defense resources, into maintaining a full complement of highly trained combat arms units and only a limited number of CS/CSS and SOF units in the active component. The remainder (approximately 75 percent in the case of CSS) of the support units reside in the reserve components (RC). Table 2.1 illustrates the Army's reliance on reserve CS and CSS units.

In the absence of a Presidential Selective Reserve Call-up (PSRC), the Army's reliance on its AC CS/CSS units to support POs raises several concerns. Many of these units are designated as part of the FSP. As mentioned above, in an MTW the Army depends on these FSP units to immediately deploy in support of the leading combat arms units. Similar RC units may be mobilized, but the AC units are the only ones that can respond in the first few days. Yet such units are being deployed to POs as well. In the event of an MTW, even if the lift assets are available to rapidly redeploy these units, their capabilities will be reduced by a lack of predeployment training, little if any time to recover and reconstitute equipment, and potentially adverse ef-

---

[22]Ritchie, Ruck, and Anderson (1994), pp. 373, 374.

[23]Sortor (1997).

fects on morale.[24] Whether they would be sufficiently prepared to go quickly into an MTW can be called into question.[25]

Nonetheless, the General Accounting Office (GAO) found that in Somalia, 50 percent of the AC support forces used were drawn from these "first-to-go" units. Specifically, 92 percent of the quartermaster forces, 69 percent of the engineering support forces, 64 percent of miscellaneous support forces, and 65 percent of transportation forces deployed to Somalia were FSP units. FSP unit types particularly stressed by the Somalia deployment included the three general supply companies, two medium truck companies (petroleum), air terminal movement control detachment, cargo transfer company, water purification detachment, and perishable subsistence team.[26] These same FSP units have also deployed in support of POs in Haiti and Bosnia, and include critical low-density capabilities such as port opening, cargo handling, and water purification and distribution.[27]

---

[24]While anecdotal evidence of decreased morale abounds, and was prevalent in our own interviews, empirical data on how multiple deployments affect soldiers remain elusive. In fact, some limited research on soldiers in Haiti appears to demonstrate that multiple rotations had no negative effects on soldiers' well-being. Wong, Bliese, and Halverson (1995), cited in Kirkland, Halverson, and Bliese (1996), p. 80.

[25]According to one study, "Army commanders generally estimate a range of 3 to 6 months to fully restore a unit's warfighting readiness after a peace operation. The 3- to 6-month recovery period is based on units' rotating or redeploying from a peace operation absent the requirement to reinforce other forces involved in a major regional conflict. Under more urgent conditions, according to DOD, the recovery period would almost certainly be shortened by freezing reassignments, curtailing leave and nonessential temporary duty, and taking other measures." U.S. General Accounting Office (1995a), pp. 34–35. Such estimates are extremely general, however, and fail to take into account, for example, the differences between unit-types' equipment and training requirements. The Army's chief planning officer in Europe is quoted in *The New York Times* as estimating of the units returning from Bosnia, "In a crisis we would still need about two months of retraining to get everyone coming back now ready." Thus, although 1st Armored Division Commander MG Nash stated "If the fate of the nation were at stake, we would be at the dock before the ships arrived," the condition of some of those forces might be in question. O'Connor (1996). Indeed, the Center for Army Lessons Learned recommended that the Army require units to report themselves at "C-5" status for a period of four months following redeployment from a peace operation. U.S. Army Center for Army Lessons Learned (1996a), cited in U.S. General Accounting Office (1997b), pp. 3–4. The issue of redeployment time from a PO to an MTW is addressed in more detail in Sortor (1997), pp. 33–46, and in U.S. Army Center for Army Lessons Learned (1995b).

[26]U.S. General Accounting Office (1995b), pp. 43–44.

[27]The GAO reports that the commander in chief (CINC) of the European Command believes that better coordination of contingency planning among the various CINCs

Table 2.1

Army Reserve Component Contribution of Selected Units to the Total Army

Unit Type	RC Contribution to Total Army Units
Water supply battalions	100%
Enemy prisoner of war brigades	100%
Civil affairs units	97%
Petroleum support battalions	92%
Public affairs units	85%
Medical brigades	85%
PSYOP units	81%
Motor battalions	78%
Hospitals	77%
Corps support groups	75%
Combat heavy engineer battalions	73%
Maintenance battalions	71%
Military police battalions	66%
Terminal battalions	50%

NOTE: Table includes Army National Guard and Army Reserve units.
SOURCE: "Reserve Component Programs," Fiscal Year 1995 Report of the Reserve Forces Policy Board, Department of the Army, March 1996, p. 12. For another view of Army Reserve contribution, see Nagy (1996), p. 14. A listing by specialty of reservists supporting Operation Joint Endeavor is given in Sullivan (1996), p. 28.

Other FSP units with multiple deployments to POs include the only active component civil affairs battalion (the 96th Civil Affairs Battalion), which has, in the absence of PSRCs, repeatedly deployed its soldiers. One author noted that of eight active Army CA specialists sent to Rwanda in August 1994, five had recently returned from other deployments to Macedonia, Namibia, Honduras, and Panama.[28] In another instance, a military police company that deployed to Somalia had been back at Fort Bragg for only four months when it was

_____

could reduce the heavy burden on these units. U.S. General Accounting Office (1996a), p. 6.

[28]Goodman (1994), p. 66.

redeployed to Haiti for Operation Uphold Democracy.[29] And although there is no documented evidence that repeated deployments affect retention, our interviews from the company level up through the FORSCOM Adjutant General suggest that, at least for midcareer soldiers with families, there is a negative impact.[30]

As detailed in a commentary for *Armed Forces Journal,* the deployment of MPs to Haiti "provides a striking example of how quickly specialized personnel resources can be strained."[31] In Operation Uphold Democracy, "about 1,500 active-duty MPs—a brigade headquarters, two battalion headquarters, and nine divisional/combat support companies—were among the operation's vanguard." At the same time, "a slightly larger MP force—a brigade headquarters, and eleven MP divisional/combat support companies—was controlling Haitian refugees at Guantanamo Naval Base, Cuba." Those two missions alone "all but exhausted the available active-duty MP assets."[32]

---

[29]Authors' interview with officers of the 19th Military Police Brigade, Fort Bragg, North Carolina, May 17, 1996.

[30]Willis (1996b). Authors' interview with 16th MP Brigade personnel, Fort Bragg, North Carolina, May 17, 1996; authors' interview with Mr. Hulett and SGM Allen, Fort McPherson, Georgia, June 24, 1996. Personnel in the 4th PSYOP Group claim that reenlistment has fallen noticeably with the frequent deployments. They point, for example, to their then-projected senior gains and losses in May 1996. They anticipated, at that time, to gain two SFCs but in the same period projected losing 2 SGs, 1 SFC, 1 CW3, 1 MSG, 2 1LTs, 1 CSM, 6 CPTs, 16 MAJs, and 2 LTCs. Data from 4th POG (A), as of May 13, 1996. On the other hand, more systematic data suggest that deployment has not been that much of a threat to retention: retention rates have held steady throughout the drawdown and concurrent increase in deployments. As with the link between deployments and morale, the link between high OPTEMPO and retention may be more a function of soldiers' expectations, which differ across types of units. For units that have not typically had high deployment rates, it may be more of a problem; however, in Special Operations units and the 82nd Airborne, among the most-deployed units in the Army, reenlistment rates have remained high. Willis (1996a).

[31]Roos (1994), p. 12. These findings were validated by the Dynamic Commitment series of wargames held in June 1997. COL Phil Coker, chief of the capabilities analysis and concepts branch under the Joint Staff's J-8 directorate and an organizer of the wargames, claimed that the wargames "shed light on the difficult issue of low density/high demand assets, or those personnel, units, or systems that are in short supply but high demand by commanders-in-chief . . . [A]ssets like military police units . . . and civil affairs units were often overtasked in the wargames." *Inside the Pentagon,* June 16, 1991, p. 1.

[32]Roos (1994), p. 12.

RELIANCE ON VOLUNTEERS LIMITS PO PERFORMANCE

In order to meet the support force requirements for POs, the Army has relied, at least in part, on RC volunteers. As an extension of this, it has experimented with forming conglomerate units composed of volunteers from different reserve units. The postal company that deployed to Somalia in 1992 and the infantry battalion that deployed to MFO duty in Egypt in 1994 are the two notable examples. Additionally, Project Standard Bearer sought to designate units of volunteers who would agree to deploy for a 45-day period if needed for any contingency.[33]

Relying on volunteers, however, whether more or less formally, can be problematic for a number of reasons. First, volunteers often fail to match the force composition requirement of the contingency. In some specialties such as civil affairs, PSYOP, linguist, and medical, it has been particularly difficult to get volunteers with the correct mix of skills and experience. Indeed, the commander of the 96th CA Battalion estimated that he can get a perfect match of skills to jobs only 30 percent of the time.[34]

Second, even if the appropriate skills are provided, it can be difficult to generate a sufficient number of volunteers, especially for MOOTW. Because Rwanda followed closely on the heels of Somalia and the amount of disease and risk of exposure to AIDS was high, the U.S. military had difficulty getting reservists to volunteer. Whereas the Army may receive as many as 1,000 phone calls from potential volunteers for a popular mission, for Rwanda the Army Personnel Center (ARPERCEN) received less than 100.[35]

Finally, the requirement for volunteers with low-density skills is high in MOOTW operations—yet many of these reservists may be reluctant to leave their civilian jobs. For example, OUD generated a backfill requirement for 48 physicians, yet only five volunteers were found. At least in part, this reluctance to volunteer can be explained by the potential economic loss: the median monthly income loss for

---

[33]U.S. General Accounting Office (1996e), pp. 8, 13.

[34]Authors' interview with 96th Civil Affairs Battalion commander, Fort Bragg, North Carolina, May 15, 1996.

[35]Brown et al. (1997).

nonphysician reserve officers during ODS was $1,765, and the median monthly income loss for physicians was $9,613.[36]

RESERVE ACTIVATION FOR POs CAN ALSO BE PROBLEMATIC

If PSRC is authorized, issues about achieving the right mix and number of reservists are somewhat mitigated. The benefits of PSRC appear to have been recognized: since Somalia, reserves were activated for both operations Uphold Democracy and Joint Endeavor,[37] enabling the Army to deliver many of the capabilities in high demand for POs. Obviously, however, such repeated call-ups can result in high RC OPTEMPO. According the Department of Defense, "throughout 1996 more than 10,000 Guard and Reserve personnel have supported Operation [Joint Endeavor], now Operation Joint Guard, from bases in Bosnia, Croatia, the U.S., Hungary, Germany, Italy and elsewhere in Europe."[38] While reserve call-ups relieve the stresses on some AC units and allow for more appropriate mixes of forces in POs, other problems result. To the extent that PO deployments are perceived as frequent events that impose a financial burden on RC personnel, reserve retention and recruiting may suffer—particularly in specialties that require a high degree of training or experience.[39] Indeed, many RC personnel hold civilian jobs that could be seriously threatened by frequent or prolonged deployments to POs. Whether the Army will be able to maintain RC retention and enlistment if these volunteers—who ostensibly sign up to serve in the

---

[36]Ibid.

[37]For a list of the numbers of Army reservists deployed to Operation Joint Endeavor by specialty, see Sullivan (1996).

[38]U.S. Department of Defense (1997).

[39]The GAO reports that "DOD has been able to obtain the reservists it needs through a combination of involuntary call-up authority and volunteerism." Reliance on volunteers, however, seems most appropriate for small-scale, relatively short-term deployments where individuals are needed to round out active units. According to the GAO, "The Army has attempted to deploy units of 50 volunteers and more, but has found that forming these size units requires 'intensive, more complex work-arounds.'" U.S. General Accounting Office (1996e).

nation's wars—are pulled away from jobs and family for more minor missions is an important question.[40]

Furthermore, policies guiding the use of reservists can affect their accessibility and even their utility. It is doubtful, for example, that PSRC could be used to mobilize low-density reservists who participated in an early rotation for a subsequent rotation, effectively capping the length of PO support for some reserve specialties.[41]

Finally, the 270-day maximum call-up under PSRC limits the number of PO rotation cycles that the RC can support, while making it difficult to maintain continuity across an operation.[42] In early 1996, one planner projected that if Operation Joint Endeavor were to continue beyond a third six-month rotation, the Army might run out of some types of reserve units.[43] When this problem actually arose, the Army responded by returning to reserve units from which soldiers had previously deployed to seek additional personnel.

This situation may be improved somewhat as the National Guard implements plans to transform 12 of its 42 combat brigades to CS and CSS units. Although the National Guard redesign planned to fill 42,700 previously unresourced support jobs, it still left 15,700 billets unfilled.[44] Moreover, in May 1997 the Quadrennial Defense Review (QDR) recommended cuts in the reserve components of 45,000; the

---

[40]Interestingly, a study of National Guard and Reserve personnel participating in MFO Sinai indicated that there is a greater appreciation in the reserve component for the legitimacy of peacekeeping missions than in the active component. Pexton (1995). It would be valuable to know whether that remains the case following the three recent activations of reservists for Desert Storm, OUD in Haiti, and IFOR in Bosnia. Katherine McIntire Peters cited some of her discussions with reservists on this topic in Peters (1996).

[41]"The Judge Advocate General and the deputy legal counsel to the Chairman, Joint Chiefs of Staff, advised that repetitive calls to active duty under PSRC for the same operation would probably be viewed by Congress as exceeding the intended scope of the PSRC statute." "Mailcall: Is It Possible to Be Mobilized Twice?" *Army Reserve Magazine,* Fall 1996, p. 28.

[42]The 270 days, moreover, do not represent feet on the ground—they also include mobilization and demobilization, as well as leave accrued over the 270-day period, which at 2.5 days per month amounts to over 22 days. Thus, assuming two-week mobilization and demobilization periods, the actual time a reservist may be operational is likely to be closer to 220 than 270 days.

[43]Authors' interview with LTC Gisler, Fort McPherson, Georgia, June 24, 1996.

[44]Graham (1996), p. 3.

degree to which these cuts will come from support units is not yet known.

CONCLUSIONS AND RECOMMENDATIONS

The challenge of both maintaining a combat-oriented force posture and meeting PO requirements, even as force structure and budgets decline, can in part be met by improving the flexibility of the existing force. Among the many possible approaches, four are already being explored and developed by the Army: greater modularity, expanded functionality, reorganizing excess assets, and increased use of capabilities outside the Army force structure.

The first approach, modularity, involves increasing the extent to which a unit can operate as an independent entity. For example, replacing the current CSB configuration of functionally specialized companies with several combined support companies—each containing transportation, supply, and maintenance elements, as well as their own command and control—would enable one or two to deploy to a PO without degrading the readiness of the others. The intention is to rewrite tables of organization and equipment (TOEs) to create integrated support elements at the company level. This would arguably give commanders the ability to better tailor their force to the scale of the mission—both in PO and MTWs—while better maintaining the readiness of nondeployed units, particularly since the stay-behinds would retain their own command and control capability. The process of implementing modularity for some types of CSS units, such as quartermaster, is currently being pursued by TRADOC's CASCOM,[45] and the current cross-training of logistics officers during their schooling facilitates this level of integration.

Modularity may go some way to meeting the wish of LTG Daniel Schroeder, commander of JTF Support Hope in Rwanda, that "units . . . be packaged in the deployment system to provide discrete

---

[45]A concern raised by some of the Army personnel we spoke with is that modular units would not provide the opportunities for enlisted leadership progression within a particular skill that now exist in the skill-specialized organizations. Although a modular unit at any given echelon should offer the same number of leadership slots as the current MTOE does, it may be the case that NCOs would have to change subunits more frequently than they now do.

capabilities rather than having to come as large organizations."[46] Yet the modular units' design is critical. If they do not precisely meet the requirements of a PO—and it would be difficult to design them to precisely meet the requirements of every operation—then force tailoring could still require taking portions of other units, obviating the benefits that modularity supposedly provides to stay-behind units. Further study is warranted to determine whether the force requirements in recent operations could have been better met by modular or functionally specialized units.

A second approach, expanding functionality, creates multirole units capable of accomplishing a broader range of tasks. To a certain extent, this has been done on an ad hoc basis: in Somalia, for example, petroleum pipeline units stood in for lower-density water distribution units. The creation of multirole bridge companies is another recent example. This is an approach that may have become more appropriate as the Army's force structure declines—units with generalized capabilities are common in the smaller engineer forces of other countries' armies.[47]

Of course, both modularity and expanded functionality have limited utility. Neither of these approaches is a blueprint for redesigning the Army; each is a response to certain force structure problems observed in recent operations.

As noted above, the Army is undertaking reorganization of its reserve components. This step was taken not because of the demands of PO deployments on CS/CSS, but because of the anticipated shortfall in CS/CSS assets if two MTWs were to take place simultaneously. Thus, the selection of support skills to which combat arms troops are converted will be dictated by the needs of the two-MTW scenario, rather than by the requirements of POs. Nonetheless, since many of the same skills are indicated—transportation and MPs, for example—the

---

[46]Schroeder (1994).

[47]Authors' interview with MAJ David W. Brinkley and CPT(P) Kelly Slaven, Fort Leonard Wood, Missouri, May 16, 1996.

reorganization will mitigate some of the effects of PO deployments on MTW readiness.[48]

A fourth and very different kind of option is to make better use of non-Army capabilities to provide support to POs. The other services, government agencies, coalition partners, the host nation, and private contractors may be able to augment or substitute for low-density Army assets, particularly those that are part of the FSP. For example, Air Force Redhorse and Navy Seabee units "saved the day" for base development in Bosnia, since the Army has kept only one combat heavy engineer battalion in USAREUR.[49] Private contractors can and do assume key responsibilities such as transportation, maintenance and repair, and constructing camps and facilities. In Haiti, the activities of Brown and Root allowed the early withdrawal of 1st COSCOM (Combat Support Command) back to Fort Bragg, thereby ensuring that the 82nd Airborne Division—which is dependent on the 1st COSCOM—would be ready if it had to deploy elsewhere.[50] Private contractors may also be particularly valuable when their ability to work outside the military logistics channels facilitates procurement of materials and replacement parts.

However, there are some problems with contractors. In some instances, we were told, private contractors are "not as responsive" as their military counterparts, requiring complicated and lengthy renegotiations when requirements change.[51] Additionally, private firms "compete for resources and infrastructure with the military," in some cases arriving late or with greater expense as they try to use the airfields or ports that are under heavy military use, for example.[52]

---

[48]There was a further proposal to reconfigure the National Guard's combat divisions from heavy to light infantry, which would make those units more flexible across MTWs and POs. The National Guard, however, refused to consider the proposal.

[49]Authors' interview with 20th Engineer Brigade, Fort Leonard Wood, Missouri, May 16, 1996.

[50]Authors' interview with Brigadier General Robert D. Shadley, USA, Director for Logistics, USACOM/J4, conducted by Dr. Wm. R. McClintock, Command Historian, at HQ USACOM, April 25, 1995.

[51]Authors' interview with officers of 20th Engineer Brigade and subordinate units, Fort Bragg, North Carolina, May 17, 1996.

[52]Authors' interview with 20th Engineer Brigade personnel, Fort Leonard Wood, Missouri, May 16, 1996.

Moreover, it is not clear whether using private contractors is a cost-effective alternative. Between October 1992 and March 1994 in Somalia, for example, the Army paid Brown and Root $77 million for water, food, transportation, fuel, and basic infrastructure support.[53]

In addition to these options, other steps could be considered. Obviously, maintaining CS/CSS units at full strength would eliminate many of the cross-leveling problems. Even if units are at close to full strength, however, the remaining shortages are likely to occur in the low-density MOSs most often cross-leveled. Identifying which MOSs have been cross-leveled most frequently could help suggest options: two possible approaches are more intensive management of those critical MOSs in terms of recruiting and retention, or enhanced incentives for reenlisting into those MOSs. Additionally, a permutation of expanded functionality at the individual level—giving soldiers multiple MOSs—could allow a unit to lose an individual without losing a critical capability. Although many soldiers already have a secondary MOS, this is usually just a residual effect of career switching upon reenlistment, and there is little expectation that the skills of the secondary MOS will ever be used. The approach contemplated here, rather, would require a focused and long-term effort to train soldiers in a low-density secondary MOS in addition to their primary MOS, and then to keep them proficient at both. Moreover, in order to be useful in this context, such a program would also require official guidance (or a system of compensation and other incentives) leading soldiers to select into the most essential specialties as their secondary MOSs and then to keep them on active duty.[54] By creating more in-unit depth in low-density MOSs, soldiers could be drawn out as individuals or in partial-unit deployments without adversely affecting the units' overall readiness.[55] Finally, making reservists in

---

[53]MacFarland (1994), p. 45.

[54]On a much smaller scale, a similar idea has been implemented in a very limited way with the combat lifesaver's program, through which selected members of a unit are trained in basic medical skills in the event of an emergency.

[55]An extreme example of this are special forces operational detachments (A-teams), which comprise a commander, a detachment technician, an operations sergeant, an assistant operations and intelligence sergeant, two weapons sergeants, two engineer sergeants, two medical sergeants, and two communications sergeants. In effect, this structure allows the team to be used as a whole, to be split into two equally capable teams, or to be task-organized to provide a team tailored to a particular mission.

the most-stressed and low-density MOSs more readily accessible, for more frequent deployments, and for longer periods (with appropriate compensation) might also be possible (although it would be challenging to develop a program that would sufficiently recompense private-practice doctors, for example, for business and clients lost during deployments). Thus far, however, DoD requests to relax call-up requirements have been repeatedly rejected.

In short, future force structure initiatives must take into account that for POs, access (to reservists and EAD/EAC forces) and the ability to tailor a force are key issues, but as long as MTW readiness remains the highest priority, those things must be accomplished without unduly reducing the readiness of the overall force.

Chapter Three

TRAINING

INTRODUCTION

As with force structure, Army training remains focused on preparation for MTWs. Indeed, most deliberate training for PO operations takes place only after units are earmarked for such duty. The inclusion or exclusion of PO and MOOTW in units' mission-essential task lists (METLs), which guide their training, remains the commanders' prerogative. Thus, although the 18th Airborne Corps has MOOTW in its METL, its subordinate units (the 82nd Airborne Division, the 101st Airborne Division, the 3rd Infantry Division, and the 10th Mountain Division—which deployed to both Somalia and Haiti) have standard combat METLs. In contrast, PO and/or MOOTW are included in USAREUR's and USARPAC's corps and subordinate units' METLs.

Likewise, given the option to conduct MOOTW exercises at the Joint Readiness Training Center (JRTC), units tend to choose the standard rotations—perhaps in part because the JRTC rotation is the focus of a battalion commander's 18-month command tour, and commanders prefer to be evaluated on what they know best. The choice of standard rotations may also stem from a concern that the units trained for POs will be the ones selected to deploy to them. USAREUR has addressed this reluctance by mandating that every rotation at the Combat Maneuver Training Center (CMTC) include a MOOTW module.[1]

---

[1]For a quick overview of the CMTC MOOTW training, see Woodberry (1996), pp. 60–61. John Woodberry is CMTC's chief of training.

This chapter examines how training primarily for combat yet deploying for POs affects units' MTW readiness and PO capabilities.

PO DEPLOYMENTS PROVIDE BOTH TRAINING BENEFITS AND COSTS

The extent to which PO deployments provide benefits to unit and individual training depends on the unit type, echelon, sequence in the rotation cycle, and nature of the operation. SOF and CS/CSS units generally perform the same missions in a PO as they would in combat, and in fact can gain invaluable experience from operating in a real-world environment (experience that is difficult or impossible to simulate at home station because of limited training budgets or environmental restrictions).[2] For example, the former commander of the combat heavy engineer battalion that deployed twice to Somalia told us that his unit did a lot more training on deployment because it was not constrained by the garrison construction budget.[3]

In combat arms, the effects are more complex. At the individual soldier level, learning how to cope with an extended deployment can teach valuable lessons. One Bradley driver, describing his experience in Bosnia, said

> we know our equipment better . . . how to keep it all going 24 hours a day . . . and we know how to pace ourselves on a mission like this one, where you have to be ready to go 24 hours a day yourself. And then with all that time working with your unit, you really get to know how to work together better than in any training exercise.[4]

Small units and their leaders may also derive some benefit from operating more independently of the platoon or company, which can both improve morale and enhance leadership abilities. For example, junior leaders pick up such skills as small unit leadership, how to keep morale up while operating in remote locations, patrol skills, lo-

---

[2]U.S. General Accounting Office (1995a), pp. 29–30.

[3]Authors' interview with LTC(P) Robert L. Davis, Fort Leonard Wood, Missouri, May 16, 1996.

[4]O'Connor (1996).

gistical resupply, and establishment of communications networks.[5] During Operation Provide Comfort in Northern Iraq, platoon leaders were challenged to respond appropriately in situations ranging from Kurdish guerrillas demanding passage through a checkpoint to the request of an Iraqi nuclear scientist who wanted to defect.[6] Army V Corps commander LTG John N. Abrams foresaw similar benefits from participation in Operation Joint Endeavor, where he believed "the lieutenants and sergeants and captains and young majors [will] learn a great deal ... They're going to develop competence and confidence that will be very important for the Army."[7]

But while training may be enhanced at the lower levels, at the company or battalion level, opportunities for collective training are impeded by restrictive rules of engagement (ROE), lack of maneuver and gunnery ranges or range access, and the operational demands of the mission. MG Joseph Kinzer, for example, said after his experience in Haiti that U.S. soldiers "lose the edge" in warfighting operations following deployment to MOOTW.[8] Army Vice Chief of Staff GEN Ronald Griffith stated that it could take as many as four months for the troops in Bosnia to recover combat proficiency lost during the peacekeeping deployment.[9] Such impressions were mirrored in many of the interviews we conducted. As one Bradley company commander from the 1st Armored Division commented when preparing to deploy home from Bosnia, his soldiers were well practiced in their PO mission, "but when it comes to attacking a position, or holding a piece of terrain against an assault, that's where we'll need work."[10] Additionally, staff combat skills may erode. MG William L. Nash, commander of the 1st Armored Division in Bosnia, has said that his staff officers, because they had been "coordinating a peacekeeping force, not planning battles," were "not as good at warfare as when they arrived." Nash went on to say, however, that "it

---

[5]Telephone interview with COL Kuenning, Army War College, February 1996.

[6]Abizaid (1993), pp. 16–17.

[7]Atkinson (1996), p. 1.

[8]"Commander of Haiti Operation Says Peacekeeping Dulls Warfighting Edge," *Inside the Army,* May 13, 1996, p. 5.

[9]Adelsberger (1996b), p. 3; "Bosnia Affecting Readiness But Army Still Able to Execute Two MRCs," *Inside the Army,* April 29, 1996, p. 2.

[10]O'Connor (1996), p. 1.

would take less than a month of practice using computer simulators to bring the skills back."[11]

Where a unit falls in the deployment schedule also affects the potential training benefits: units that deploy at the beginning of a PO accrue experience in planning for and operating in an unstable and unpredictable environment, while follow-on unit planning and activities may be much more routinized. For medical units, for example, the training value is

> highest during the initial rotation and the first month or so of a rotation, [and] is often limited to the initial phase of a deployment (i.e., experience in establishing a field hospital) . . . [A]s a result, some medical personnel may be underutilized which offsets the training value gained in other areas.[12]

Military police officers who deployed to OUD shared this experience, claiming that only "the first month or so" of the operation provided them any training benefit.[13]

Finally, opportunities for combat training depend heavily on the nature of the PO. For Task Force 2-87, on the third day of its deployment to Somalia,

> rioting and street fighting erupted throughout Kismayu, and Task Force (TF) 2-87 deployed security forces to track down gunmen and terrorists who were randomly spraying civilians with small-arms fire and grenades. During this operational phase, TF 2-87 elements killed five gunmen, wounded or captured five others and apprehended 12 bandits with no friendly losses.

In addition,

> TF 2-87 medics treated more than 70 Somalis, many of whom had been gruesomely tortured by rival factions.[14]

---

[11]Ibid.

[12]Davis, Hosek, Tate, et al. (1997).

[13]Authors' interview with officers of the 16th Military Police Brigade, Fort Bragg, North Carolina, May 17, 1996.

[14]Stanton (1994), p. 35. Another account of close combat in Somalia appears in U.S. Army Center for Army Lessons Learned (1993–1994), p. I-4-12. More U.S. soldiers were

GEN George Joulwan noted similar combat training benefits, stating that the exacting demands of that operation had brought about an overall improvement in the Army's readiness and combat capability.[15]

Units in Macedonia for Operation Able Sentry, in contrast, are constrained to manning static observation posts and thus have little opportunity within the conduct of their mission to maintain combat proficiency. One mechanized infantry unit that deployed to Macedonia in 1994 received the lowest score in its divisionwide Bradley qualification test upon redeployment. This poor performance was attributed to having deployed without its primary tactical vehicle, the Bradley Fighting Vehicle, not having access to a Bradley simulator, and UN guidelines that barred it from engaging in collective training while in Macedonia.[16]

PO DEPLOYMENTS INTERRUPT TRAINING CYCLES

While experience during a PO deployment may or may not yield training benefits to a unit, the increased OPTEMPO resulting from participation in POs inevitably disrupts units' regular training cycles and often requires the use of operations and maintenance (O&M) monies originally intended to fund training. Thus, the 3rd Battalion, 73rd Armor Regiment had to cancel its annual gunnery evaluation when it deployed to Haiti. Similarly, an aviation task force from 18th Aviation Brigade returned from Haiti "untrained for wartime requirements," in part due to lack of night flying during deployment.[17] In some cases, units have barely had time to recover from one PO before being deployed to another: the 10th Mountain Division's aviation brigade, for example, deployed to Haiti in Operation Uphold Democracy less than six months after returning from more than a year of operations in Somalia.[18] Because combat training progresses

---

killed in Somalia than during Operation Just Cause in Panama in 1989, and the per-capita casualty rate was higher in Somalia than in Desert Storm. Kirkland, Halverson, and Bliese (1996), p. 82.

[15]Breen (1996), p. 85.

[16]U.S. General Accounting Office (1995a), p. 29.

[17]Shelton memo (1995).

[18]Archambault (1995).

from individual skills to small-unit tasks to large-unit collective training, these disruptions and associated personnel turnover often mean that units spend less time on collective training, thereby reducing MTW readiness.

UNMET PO TRAINING REQUIREMENTS CAN UNDERMINE PO EFFORTS

While deployments to POs affect units' readiness, these operations also have their own requirements, although these are more related to conditions and standards than to unique PO tasks. Indeed, most PO and MTW tasks overlap—for example, identifying belligerents and equipment, preparing and occupying an observation post, and conducting countermine operations are all tasks common to both types of deployments.[19] There are some tasks unique to POs, such as mediation and negotiation, but these make up a relatively short list.[20] Yet although most tasks are the same, the conditions and standards to which units must train will differ based on the nature of the mission. In POs, for example, patrolling is overt rather than covert, and reacting to hostile contact requires increased consideration of

---

[19]For a discussion of training for peacekeeping and peace enforcement, see Stennett and Walley (1993).

[20]Whether or not these few additional tasks impose requirements for additional training is debated. Some studies, such as that by Wiseman (1983, p. 183), find that units trained well for combat "have little difficulty adapting to peacekeeping . . ." In comparison, MAJ Phillip Brinkley, in his study of the tactical requirements for peace-keeping, determined that "soldiers need intensive training to enhance their tactical military skills." Brinkley (1985), p. 35. Surveys of soldiers participating in peace opera-tions can also have opposing results. One survey of U.S. peacekeepers in MFO Sinai, for example, found that more than 80 percent of the soldiers, across three units, felt that they needed PO-specific training in addition to their combat training. Segal, Harris, Rothberg, and Marlowe (1984), p. 495. Another survey by the same sociologist, of soldiers deployed to Haiti, found that 60 percent of the 10th Mountain Division (Light Infantry) soldiers believed that professional soldiers were able to perform peacekeeping and combat missions equally effectively (although a large majority of the division's personnel also believed that they needed additional peacekeeping training to augment their training in basic military skills). Pexton (1995). LTG Henry Shelton, commander of Joint Task Force 180 in Haiti, cited the negotiations training he received during a two-week Harvard course on security environments as extremely valuable preparation for his efforts in Haiti. Shelton Interview Transcript, Automated Historical Archives System (AHAS), Fort Leavenworth, February 13, 1995.

collateral damage and civilian casualties.[21] The commander of the 3rd Squadron, 2nd Armored Cavalry Regiment in Haiti described how soldiers faced with potentially threatening crowds nonetheless kept their weapons slung, and he emphasized that in peacekeeping—as opposed to peace enforcement or combat—some "deprogramming" of typical soldier skills is required.[22] One soldier from the 1st Armored Division in Bosnia explained that

> our training was to maneuver and take the enemy out. Here we've had to learn a different concept. We had to learn not to shoot because you don't really know who your enemy is. You have to sit back, watch and try to keep the peace.[23]

POs' more restrictive rules of engagement present "additional parameters or conditions within which units must operate to accomplish peace enforcement tasks."[24]

These distinctions between MTW and PO conditions and standards are nontrivial, and soldiers require education, training, and exercises to prepare them adequately for the often frustrating, confusing, and changeable PO environments. As MG William L. Nash described it,

> I've trained for 30 years to read a battlefield. Now you're asking me to read a peace field. It doesn't come easy. It ain't natural; it ain't intuitive. They don't teach this stuff at Leavenworth . . . It's an inner ear problem. No one feels completely balanced.[25]

---

[21]As reported by the GAO, "in Haiti the night patrols were conducted under full illumination, as a show of presence, rather than in a more stealthy manner, as is the case in war." U.S. General Accounting Office (1995a), p. 16. For a sample Task and Evaluation Outline for a peace operation task, "Cordon and Search," not currently included in Army doctrine, see U.S. Army Center for Army Lessons Learned (1993–1994), p. I-4-10.

[22]Niblack (1995), p. 30.

[23]Palumbo (1996), p. 17.

[24]U.S. Army Center for Army Lessons Learned (1993–1994), p. I-4-5.

[25]Atkinson (1996), p. 1. A Canadian soldier involved in the United Nations Truce Supervision Organization (UNTSO) had a similar observation: "Finally, one was reduced to two necessary human characteristics: patience and understanding, plus any amount of tact and diplomacy! As a general rule, these latter traits are not always found in military men." *Peacekeeping*, Canadian Forces Command and Staff College, Command and Staff Course 4, 1978, p. F-1, cited in Brinkley (1985), p. 11.

Adding new capabilities like nonlethal weapons makes this even more of a challenge—U.S. personnel in Somalia were issued cayenne pepper spray as

> an effective means of proportionate force against low-level threats . . . In one instance, a Somali attacked a soldier with a knife. Instead of shooting the Somali, other soldiers nearby used the spray. Although the spray worked and the Marine escaped un-harmed, the Somali had attempted to stab one Marine four times before he was subdued with the spray. In this case, deadly force may have been called for, but the Marines saw cayenne pepper spray as a substitute for deadly force, instead of as a complement.[26]

To better train his junior officers to cope with the unique judgmental demands of PO, one brigade commander with whom we spoke told us that he conducts situational training exercises during officer pro-fessional development time. He also told us, however, that what he really needed was a situational training exercise range to train his soldiers on ROE-related skills that were not part of their Army school training.[27]

UNMET PO TRAINING REQUIREMENTS RESULT IN LOWER MORALE

Not only must soldiers be prepared to make extremely difficult judgments quickly (e.g., "Am I being threatened?" and "What's the appropriate level of response?"),[28] but, as discussed in a series of articles published by the National Center for Post-Traumatic Stress Disorder, they must also be prepared for the unusual powerlessness they may feel in a PO.[29] It is not uncommon for soldiers in POs, for

---

[26]Dworken (1994), p. 31.

[27]Authors' interview with COL Thompson, commander of the 20th Engineer Brigade, Fort Bragg, North Carolina, May 17, 1996.

[28]Indeed, one of the USAREUR Lessons Learned from training at Mountain Eagle 95 was that peace enforcement operations (PEO) "require extensive discretionary decision training for the individual soldier and small units such as platoons, squads, crews, and teams." *Peace Enforcement Operations,* Lessons Learned Newsletter #1-95, MOUNTAIN EAGLE 95, USAREUR, p. 5.

[29]Of course, this may not be universally true, and will depend on the nature of the operation. Those operations more akin to traditional peacekeeping, in which the

example, to witness violence to civilians or other criminal activities but be prohibited from responding.[30] This was the case, for example, in Haiti, where U.S. soldiers interviewed by RAND staff expressed frustration that they could not intervene in local disputes except in cases where grievous bodily harm was likely to result.[31] Relatedly, medical staff of the 86th Evacuation Hospital were frustrated because they were prohibited from treating local Somalis.[32]

Although it has become almost a truism, it is worth observing that the best possible preparation for the challenging PO environment is discipline. Highly trained and professional combat soldiers may be the most likely to experience frustration and boredom in what can essentially be policing duties during peace operations,[33] but they have proved to be the best soldiers for the job.[34] Nonetheless, in addition to discipline, soldiers can benefit from predeployment training and exercises in which the reasons for the limits on their activities are made clear: for example, feeding candy bars to starving children can be deadly; administering aid to the local population can create impressions of favoritism and can also raise local expectations of further such treatment to unrealistic levels; interfering in civilian violence can delegitimize the local police; and so forth. Such infor-

---

soldiers have been invited in and a cease-fire is in place, may be much less stressful—and more rewarding—than more ambiguous operations in more dangerous environments. Even so, Laura Miller found in her research on soldiers participating in Able Sentry—a relatively traditional peacekeeping operation—that "[s]oldiers who had served a full six months and who were on the patrolling teams were demoralized because even within the parameters of peacekeeping, they appeared ineffectual." Miller (1997), p. 440.

[30]Weisaeth, Aarhaug, Mehlum, and Larsen (1993), cited in Litz (1996), p. 4; Kinzer (1995), pp. 1, 12; and Weisaeth, Mehlum, and Mortensen (1996), p. 13.

[31]Niblack (1995), p. 29.

[32]Ritchie, Ruck, and Anderson (1994), p. 374; Adelsberger (1996a), p. 20.

[33]One study found that combat-trained U.S. light infantry soldiers were not optimally suited for MFO Sinai, since they tended to become frustrated. Segal, Furukawa, and Lindh (1990). A discussion of the leadership challenges imposed by POs can be found in U.S. Army Center for Army Lessons Learned (1993–1994), p. I-4-19.

[34]Segal and Segal (1993), pp. 56–69, cited in Litz (1996). Observers at the U.S. Combat Maneuver Training Center found that "well trained units and soldiers can readily adapt and become highly proficient at peace enforcement operations [as long as they retain] METL-based, performance-oriented training, adapting it to the different conditions found in PEO as necessary." *Peace Enforcement Operations*, Lessons Learned Newsletter #1-95, MOUNTAIN EAGLE 95, USAREUR, p. 2.

mation in advance can help soldiers make correct decisions and feel confident that their inaction, when called for, is appropriate.

ARMY RELIES ON JUST-IN-TIME PO TRAINING

The Army's approach thus far, consistent with its emphasis on combat training, has been to offer "just-in-time" preparation for units designated to deploy to POs.[35] Such an approach has worked relatively well, in fact. In its 1994 study on how the U.S. military services conduct PO training and education, Science Applications International Corporation (SAIC) describes the Army's policy as providing specialized intensive training for units once they have been assigned a peace operations mission, at which point they can request and use Peace Operations Training Support Packages for home station training and even deploy to the appropriate combat training center (CTC) to conduct a peace operations exercise before deployment, if time permits. Mobile training teams (MTTs) can also be dispatched to the deploying unit to prepare it for the specific requirements of the contingency.[36] This "last-minute" policy only works, of course, if sufficient time is available prior to deployment for the units to take advantage of the training packages and exercise opportunities. Even if there is sufficient time, moreover, training may continue to be underemphasized: a June 1995 interim CALL report found that units designated to deploy to POs spend most of their available time executing their standard operating procedures (SOP) for deployment and little time on specialty training for the mission.[37]

Whether last-minute or more routine, recent experience has demonstrated that PO training can benefit from the establishment of an institutional memory upon which deploying units can rely for guidance and information. For example, MG Montgomery Meigs, commander of the 1st Infantry Division, which replaced the 1st

---

[35]For example, before deployment to Operation Provide Comfort in Kurdistan, 3rd Battalion, 325th Airborne Battalion Combat Team developed and conducted checkpoint drills, countermine training, and wargames for platoon and squad leaders to help them prepare for the sensitive political environment. Abizaid (1993), pp. 12–13.

[36]Science Applications International (1994), pp. 44–45.

[37]The CALL report is cited in U.S. General Accounting Office (1995a), p. 26.

Armored Division in Bosnia, highlighted the use to which his soldiers put lessons learned by their predecessors:

> We had more time to prepare than 1st Armored Division did last year . . . more time to develop scenarios with role players and type of actual situations that confronted 1st Armored and our . . . allies. We took the same approach in the Command Post Exercise for the headquarters . . . in which all the "friction of war" one could expect in the actual area of operations was salted into four and a half days of continuous operations . . . designed to put all our leaders, includ-ing me, in the pressure cooker one can expect of that part of the world.[38]

In a similar vein, Meigs said in an earlier interview that

> We've learned the value of deliberate preparation and emphasized it. Thanks to lessons learned from the 1st Armored Division, we've gained a better respect for the requirement for documentation throughout the mission, and we've picked up a bunch of techniques and procedures they pioneered that will be extremely useful. The operational insights we've gained from the IFOR force have been super.[39]

PARTIAL UNIT DEPLOYMENTS IMPAIR PO PREDEPLOYMENT TRAINING, INTERRUPT COMBAT TRAINING CYCLES

As mentioned in Chapter Two, partial unit deployments can also have a negative impact on the stay-behind units' training capabili-ties. Rear detachments may lose key people or equipment that are critical to the continuation of normal unit training. This has been true of the active PSYOP battalions, for example, which have lost key leaders to deployments, as well as for the 20th Engineer Brigade, which has seen its topographic companies reduced to 65 percent strength due to deployment of elements to Bosnia. Where the de-ployed individuals are low density and already in short supply, as with this unit's topographic surveyors and terrain analysts, this may

---

[38]"New Mission in Bosnia Has Same Dangers," *USA Today,* October 22, 1996, p. 11.

[39]Barham (1996), p. 4.

mean that the remaining elements may be too small to conduct meaningful training.[40]

Uncertainty in the amount and timing of reimbursement for unit operating funds spent on POs also may inhibit rear detachment training. This occurs because deployed units often must pay expenses of the deployment out of their unit operating funds—the combat heavy engineer battalion in Somalia spent operating funds to construct its base camp, for example—but then may not be fully reimbursed.[41] Furthermore, rear detachments tend to focus their activities on supporting the deployed element as a replacement base and processing activity, rather than on conducting their own training.[42] The cumulative effect is a degradation in rear detachments' overall training.

CROSS-LEVELING INTERRUPTS COMBAT TRAINING CYCLES, IMPAIRS PO PREDEPLOYMENT TRAINING

Unit disruption from cross-leveling can also interrupt combat training cycles by drawing key personnel from rear detachments to supplement deploying units. Cross-leveling also affects training for the PO, since the augmentees frequently have not conducted any predeployment training as part of the unit with which they are actually deploying (and may not have conducted any predeployment training at all).

CONCLUSIONS AND RECOMMENDATIONS

To develop PO skills, training must, among other things, enhance soldiers' understanding of the difference between PO and MTWs; recondition soldiers from massive response to minimum force; and prepare soldiers for the unique joint, combined, interagency PO en-

---

[40]Authors' interviews with 1st PSYOP Battalion personnel, Fort Bragg, North Carolina, May 15, 1996, and 20th Engineer Brigade personnel, Fort Leonard Wood, Missouri, May 17, 1996.

[41]Authors' interview with officers of the 20th Engineer Brigade, Fort Leonard Wood, Missouri, May 17, 1996. Even if they are fully reimbursed, they may have missed their training opportunity and been unable to reschedule.

[42]Authors' interview with COL Thompson, Fort Bragg, North Carolina, May 17, 1996.

vironment (and the additional challenge of working closely with local officials, nongovernmental organizations, and UN representatives). To accomplish those things while also maintaining MTW readiness is the challenge; there are no obvious win-win solutions. Time spent training for one mission is time lost for training on another. As the Army has recognized, the keys to creating soldiers who can operate flexibly across both environments are discipline, well-educated commanders, and intensive predeployment and redeployment training and exercise programs.

Because most PO and MTW tasks overlap, POs call for a different training emphasis, rather than training on different tasks. Indeed, perhaps the most unique PO training requirement is for education and exercises that address the tension that neutrality and restraint can engender in combat-trained soldiers placed in volatile and dangerous conditions.

Various efforts along these lines are already available at the leader, staff, soldier, and unit levels. Officer and NCO schools are a prime opportunity for educating leaders on the basics of POs. The U.S. Army War College's core curriculum addresses POs and includes case studies, and this is complemented by three elective courses on "Collective Security and Peacekeeping," "Peace Operations Exercise and Conflict Resolution," and "Strategic Negotiation" that focus specifically on peace operations. Additionally, the U.S. Army Command and General Staff College teaches peace operations as part of its OOTW course.[43]

Staff training exercises at battalion level and above can highlight the differences between planning and executing a PO and a combat operation. In Haiti, for example, MG Kinzer insisted on conducting a two-week staff training exercise to develop staff skills and facilitate team building. He modeled the exercise after the Battle Command Staff Program at Fort Leavenworth and invited personnel from that program, as well as experienced French, Canadian, and Nordic peacekeepers, to help conduct the exercise.

At the squad level, situational training exercises can help soldiers learn to respond appropriately under more restrictive ROE, and for

---

[43]Boyd (1995), p. 29.

larger units, PO-based JRTC and CMTC rotations can be beneficial for exercising and refining a unit's PO skill set.

With the exception of the JRTC and CMTC PO rotations, none of these education or training efforts demands trading off MTW training for PO training. In contrast, the limited training opportunities during deployments can be used—at unit commanders' discretion—either for retaining MTW capabilities or for further developing PO skills, but not for both. Thus, some commanders have focused on maintaining their units' combat skills, conducting individual and small-unit combat training either at the PO sites or elsewhere in theater. For example, in Haiti, 10th Mountain Division and 25th Infantry Division units took advantage of a former Haitian military firing range.[44] U.S. forces also set up rifle, helicopter, and artillery ranges in Bosnia, and tank and Bradley gunnery ranges in Hungary[45] for the use of combat units assigned to Operation Joint Endeavor.[46]

Even during deployments where extensive ranges and other training facilities may not be available, soldiers can conduct some useful combat training. Programs like the expert infantryman badge and expert field medical badge tests emphasize skills that can be practiced at the squad and team level, even while soldiers are on duty at isolated observation posts or checkpoints.[47] Other unit leaders have instead focused on improving capabilities for the mission at hand. For example, rather than training for MTW tasks, some U.S. Army units supporting UNOSOM II in Somalia set up MOUT facilities and ranges in country to practice building and room clearing. Determining that the standard battle drill was often inappropriate for the ROE, these units adopted the special operations forces' close-quarter combat technique, which "coupled with the use of 'stun' or 'stinger' grenades, saved lives, reduced needless expenditures of ammunition, and minimized collateral damage." Which is more ap-

---

[44]U.S. General Accounting Office (1995a), p. 33; Caldwell (1995).

[45]O'Connor (1996).

[46]Conducting normal sustainment training so that unused skills would not atrophy was a lesson that the British learned during Operation Grapple in Bosnia and Herzegovina and reported through the Doctrine Branch, Director General Land Warfare, within the UK Ministry of Defense. Cited in USAREUR Lessons Learned, *Operation Grapple,* Summary of Lessons Learned Analysis, December 1995.

[47]U.S. Army Center for Army Lessons Learned (1993–1994), pp. I-4-7, I-4-12.

propriate—focusing on PO or MTW skills—will depend on the unit type, the nature of the mission, and so forth.

While the Army has made strides in providing PO training, some of its efforts could be enhanced or made more systematic. MG Kinzer's staff-training exercise prior to Haiti, for example, was deemed successful: it could be used as a blueprint for future operations. It seems apparent, in contrast, that the predeployment training for Bosnia—both in CONUS and in Europe—was less successful, at least initially: soldiers who went through it reported that it was inefficient, that critical PO preparation (mine-awareness training and orientation on Bosnian politics and culture, for example) was inadequate, and that it was oriented to refreshing basic soldier skills as opposed to offering mission-specific skills.[48] Examining how this training program was set up and how it changed over time, and assessing what worked and what didn't, could help in developing standard operating procedures that would streamline future PO predeployment training, making it more effective and less time consuming.

Finally, there may be no adequate solution to the problem of diminished collective training opportunities for units deployed to POs. Staff training exercises and simulations may help maintain combat proficiency at the command and staff levels, but we have not identified any substitutes for company- and battalion-level live-fire combat training. Nor can the effects of partial deployments on rear detachments' collective training be readily mitigated in a resource-constrained environment.

---

[48]Authors' interviews with MAJ Barge, CALL team member, Fort Leonard Wood, Missouri, May 16, 1996, and discussions with RAND and Center for Naval Analyses analysts who participated in the 7th ATC training course.

Chapter Four

EQUIPMENT ACQUISITION, SUPPLY, AND MAINTENANCE

Just as the Army's combat focus determines current force structure and training, this focus is evident in the distribution of equipment across echelons and by unit type, as well as in terms of how much of what kinds of equipment is available. This chapter examines how this focus, combined with frequent deployments to POs, affects both PO and MTW equipment capabilities.

PO DEPLOYMENTS CAN ERODE MTW EQUIPMENT READINESS

Wear and Tear on Equipment

Long field deployments coupled with high equipment usage rates[1] and limited maintenance capabilities create conditions under which equipment deteriorates more rapidly than in typical home-station training, and the nature of the damage may require longer repair periods.[2] Some units returning from POs have therefore taken longer to recover their equipment to full deployment readiness than might be expected.[3] Also, use of equipment inappropriate to the rugged

---

[1]The tempo of operations of the AH-1F and UH-60 aircraft assigned to the aviation task force supporting UNOSOM II was over 400 percent and 300 percent of the norm. U.S. Army Center for Army Lessons Learned (1993–1994), p. I-5-11.

[2]Authors' interview with officers of 16th Military Police Brigade, Fort Bragg, North Carolina, May 17, 1996.

[3]"Upon their return from Somalia, the 10th Mountain Division's AH-60 helicopters had to enter depot level maintenance as a result of the harsh desert environment and the extensive use of these helicopters in Somalia." U.S. General Accounting Office

conditions found in many POs has resulted in significant damage. The M915 tractor, assigned to echelons above corps (EAC) medium truck companies, was used extensively in Somalia to haul cargo on unimproved dirt roads. But the M915 is not designed for that type of terrain, and its operation readiness rate at times dipped "as low as 22 percent."[4]

Equipment Left Behind

In several cases, redeploying units have been directed to leave key items of equipment in theater for use by follow-on Army units, civilian contractors, or the United Nations. In Somalia this included engineer, medical, and water purification equipment, and the equipment of the active Army's only nonperishable subsistence unit.[5] Many of these items have not been replaced in a timely manner, preventing the unit from conducting training and ultimately reducing its MTW deployability.[6]

Equipment Modified

Some units modify their equipment to meet the demands of the operating environment, to address concerns about interoperability, or to comply with rules of engagement.[7] In such cases, the assumption that units can rapidly redeploy from a PO to an MTW must take into account the delay caused by restoring equipment to a combat-ready configuration. Although some restoration can be done prior to redeployment from the PO (repainting vehicles, for example), other

---

(1995b), p. 48. In our unit interviews, engineers in particular seemed to have experienced these kinds of problems.

[4]U.S. Army Center for Army Lessons Learned (1993–1994), p. II-10-29.

[5]"Impacts of Operations Other Than War (OOTW) on Unit Readiness," *Memorandum for Commander, XVIII Airborne Corps,* February 1995.

[6]In U.S. General Accounting Office (1994), the GAO reports that although over $44.2 million of equipment items was sold to the United Nations during the transition from Operation Restore Hope to UNOSOM II, "Prior to the sale and lease of these items, the Army, which owned most of the items, studied the impact of these transactions on unit readiness and concluded that they would not lessen unit readiness." Shelton memo (1995), pp. 5–6. Based on our interviews and literature review, unit readiness was impacted in at least several cases.

[7]U.S. Army Center for Army Lessons Learned (1993–1994), p. I-4-20.

modifications require more significant effort. MPs from the 18th Airborne Corps who deployed to Somalia replaced the frequency-hopping radios in their vehicles with an older radio type to be compatible with other deployed units. Should these MPs have been called upon to redeploy to an MTW, their organic radios would have had to be shipped from Fort Bragg for reinstallation.[8] In another instance, one infantry battalion modified its HMMWVs so it could use them as troop carriers; again, redeployment to an MTW would have required additional time to redress the change.[9]

Some Positive Effects

Although PO deployments can inhibit equipment readiness for combat, they can also offer some benefits. In an interview with *Defense Daily*, MG Nash discussed how the Bosnia mission had facilitated real-world testing of new technologies. According to Nash, "we are well down the highway . . . [Bosnia] has taken us from the tank trail to the Autobahn in automation." The general highlighted the Army's use of unmanned aerial vehicles (UAVs) for real-time intelligence gathering and a computer-generated logistics tracking system, Total Asset Visibility (TAV), which seeks to improve the speed and efficiency of supply operations.[10] Another example is the use of the Army's AH-64 Apache helicopter as a surveillance platform. According to a pilot attached to the 2nd Battalion, 227th Aviation Regiment in Bosnia, "what we have done is validate the helicopter's ability to gather information . . . ." Another officer added, "we have developed new [tactics, techniques, and procedures] for the Apaches here; namely reconnaissance."[11]

---

[8]Authors' interview with officers of the 16th Military Police Brigade, Fort Bragg, North Carolina, May 17, 1996.

[9]In Kurdistan, the 3rd Battalion, 325th Airborne Battalion Combat Team needed to convert its normal carriers into troop carriers; since conversion kits were not readily available, the battalion's mechanics built facsimiles locally. The same unit provided additional squad automatic weapons and M60 machine guns to its TOW missile platoons, but had to mount them alongside the TOWs. Abizaid (1993), p. 14.

[10]Bender (1996b), p. 85.

[11]Bender (1996a), p. 96.

CROSS-LEVELING EQUIPMENT AND PARTIAL UNIT DEPLOYMENT CAN WEAKEN STAY-BEHINDS' READINESS

Deployments frequently require the cross-leveling of equipment. Although such cross-leveling allows the unit deploying to the PO to be fully mission capable, the stay-behinds are either short the equipment signed over to the deploying unit or have traded for unserviceable equipment. To the extent that the stay-behind unit lacks the serviceable equipment required for its mission, its equipment readiness and deployability in the event of an MTW is decreased.

Partial unit deployments can also reduce the equipment readiness of the stay-behind element if the subunits are dependent on each other for certain types of specialized equipment. Engineer and signal units fall into this category, but medical units are particularly vulnerable due to their low equipment densities: for field hospitals, for example, "if a third of the unit is deployed, the equipment requirement to support it could entail sending the hospital's only complete x-ray, central material service (CMS), pharmacy, operating room, blood bank, laboratory, medical maintenance, or occupational therapy/physical therapy sections."[12]

MAKE-DO EQUIPPING CAN HARM BOTH PO PERFORMANCE AND MTW READINESS

Despite such problems, deployed units have focused on doing what they need to do to accomplish the mission, and they have innovated with the equipment on hand—for example, infantry units in Haiti used mine detectors as metal detectors to search visitors to the U.S.-controlled facilities[13] and, in Somalia, the absence of armored, wheeled vehicles for patrolling led Army personnel to load HMMWVs with sandbags for protection.[14] Slow replacement of equipment and spare parts has also encouraged cannibalization of equipment in

---

[12]Brown et al. (1997).

[13]U.S. Army Center for Army Lessons Learned (1994), p. 131.

[14]U.S. Army Center for Army Lessons Learned (1993–1994), p. I-4-20.

theater, which can only be a short-term solution.[15] Some units do not wait to find themselves short once in theater: one MP unit in Haiti anticipated supply lags by deploying with excess prescribed load list (PLL) items.[16]

Equipment issues also influence how commanders in theater choose to employ their limited resources. As discussed in a CALL report from Haiti, "there was a reluctance to replace MP forces [in static missions] with less mobile, non-MP forces due partly to the imposing presence of the MP team's crew served weapon mounted on the hardtop HMMWV. These teams were generally parked in and around the guarded facility." Tying MPs to buildings meant that they were unable to conduct other METL tasks effectively, such as conducting counterlooting patrols and other public safety missions.[17] In another instance in Somalia,

> tube-launched, optically tracked, wire-guided [TOW missile] vehicles were useful in the cordon areas outside of population centers where intervisibility distances were greater. They used their thermal sights to pick up movement and had the mobility to check out any suspicious activity.[18]

Although using equipment in ways for which it was not designed, creating "motor-pool queens" through cannibalization, or depleting unit funds and installation stocks through hoarding address some of the immediate maintenance and supply problems faced by deployed units, they are obviously suboptimal solutions for POs and adversely affect MTW readiness.

---

[15]U.S. Army Center for Army Lessons Learned (1993–1994), p. II-10-39 and p. II-10-47, gives examples of cannibalization by engineer and water-purification units because of lack of repair parts.

[16]Authors' interview with officers of the 16th Military Police Brigade, Fort Bragg, North Carolina, May 17, 1996.

[17]U.S. Army Center for Army Lessons Learned (1994), pp. 142–143.

[18]Stanton (1994), p. 38.

POs' EQUIPMENT REQUIREMENTS DIFFER FROM MTWs'

POs do not require much unique equipment, but they do require a different distribution than MTWs across unit types and echelons, as well as different quantities of specific equipment types. Vehicles, which are essential for quick reaction forces, routine patrolling, and logistics, are in short supply in the light infantry divisions that were deployed to Somalia and Haiti. Similarly, light divisions only have an organic engineer battalion, rather than an entire brigade with its full complement of engineer equipment, as would be the case in a heavy division.[19] Finally, while the Army has crowd- and riot-control gear, it is not widely distributed nor readily available in large enough quantities.

Other equipment issues in POs include supply and maintenance, which have been problematic in recent operations. Planners who build TPFDDs and the CINC staffs who approve them have not always taken into account such requirements as early arrival of PSYOP gear and Class IV building materials. In Somalia, for example, the construction of base camps was impeded by a shortage of construction materials.[20] Additionally, forces for POs tend to be more highly tailored than for traditional missions, so standard supply and maintenance assets may not be up to supporting the diversity of attachments, which include support of nongovernmental organizations (NGOs), other agencies, and coalition partners.

Ad Hoc Approach to Supply Exacerbates PO Maintenance Challenges

Equipment maintenance problems for units deployed to POs have been further complicated by an ad hoc approach to supply. In Somalia, "there were six separate and distinct supply support systems in use." Further, the lack of centralized theater management prevented any cross-leveling of supplies or equipment in theater. "There was no viable visibility of due-ins with this number of systems in place," and "no off-line management of requests coded with ex-

---

[19]Authors' interview with LTC Whiteman, S-3 of 41st Engineer Battalion during ORH, Fort Leonard Wood, Missouri, May 16, 1996.

[20]Zvijac and McGrady (1994), p. 56.

ceptional data above the DS level. These circumstances led to the buildup of 'iron mountains' of repair parts and materiel accumulated through redundant requisitioning."[21]

Bosnia provides a similar example: due to the high political profile of the mission, the Class IX management system for Operation Joint Endeavor is forward-based, large, and redundant, with the standard division support operations and division materiel management center (DMMC) augmented by a nondoctrinal corps materiel management support team (CMMT).[22] In addition to these support elements, there are also staff elements from a number of organizations located at the intermediate staging/support base (ISB) in Hungary. According to the CALL Initial Impressions report, "this layering of support often causes a crossing of 'lanes' and significant duplication of effort," and having all these staffs deployed forward may have "provided little value added."[23] Again, it seems as if the ad hoc approach fails to tie together activities that would be more fully coordinated in the event of an MTW.

The more general complaint we heard of slow resupply seems to be based on a variety of problems—physical transmission of requisitions at the unit level, batch processing at the supply support activity, delays in building pallets and filling trucks, and bottlenecks in the air lines of communication (ALOC). These are issues that are probably not unique to POs, and the Army is actively addressing them through its Velocity Management initiatives. To the extent that they are frustrated through normal channels, commanders have the option of purchasing local materiel and equipment that already exists in the Army inventory. In Haiti, for example, class A agents (individuals authorized by the commander to make local purchases) made a trip to the Dominican Republic and Miami, Florida, to obtain repair parts.

---

[21]The six systems were modified DS4, home station reliance, direct telephone to NICPs/depots, UN supply system, higher headquarters or senior officer intervention, and the AMC system. For a more detailed explanation of these systems, see U.S. Army Center for Army Lessons Learned (1993–1994), p. II-10-21.

[22]The high level of support may not have been immediately effective: one of the CALL team members told us that early in the mission, logisticians responded to delays by creating an ad hoc requisition code to expedite resupply for deployed units. Authors' interview with MAJ Barge, CALL team member, Fort Leonard Wood, Missouri, May 16, 1996.

[23]U.S. Army Center for Army Lessons Learned (1996b), p. 22.

One resupply problem we heard about often in our interviews with engineers appears to stem primarily from the low-density nature of these units, rather than systemic deficiencies in the supply organization—combat heavy engineer units predictably had difficulty acquiring replacement parts for their heavy earth-moving equipment in both Somalia and Haiti.[24] Because division-level forward support battalions (FSBs) do not normally stock parts for EAD units, this type of difficulty should be anticipated for corps-level attachments, particularly engineers, aviation, artillery, air defense, and signal units, and resupply channels should be identified before deployment.

Finally, some supply complaints may have to do with mission-unique support arrangements with outside contractors. In Somalia, for example, one arriving MP unit took over vehicles used by the previous unit but then had a difficult time obtaining spare parts resupply for those vehicles from the LOGCAP contractor, Brown and Root.[25]

Force Tailoring Can Create Supply and Maintenance Problems in POs

Some units have deployed to POs with non-TOE equipment that their personnel are not trained to maintain, and other units have been attached to headquarters that are not prepared to maintain or supply their TOE equipment. For example, the 43rd Engineer Battalion reported difficulty in obtaining repair parts in Somalia because the unit to which it was attached, the 10th Mountain Division, does not have an organic combat heavy engineer battalion.[26]

---

[24]On D+25, the engineers still had not received any Class IX repair parts through the normal system. U.S. Army Center for Army Lessons Learned (1994), p. 208.

[25]Authors' interview with officers of the 16th Military Police Brigade, Fort Bragg, North Carolina, May 17, 1997.

[26]Similarly, the 10th Mountain Division's 2-87th relied on the Marines for logistical support in Baledogle in December 1992. Four months later, the logistical system, since transferred to the Army, remained unresponsive to certain requirements, such as self-service supply centers (SSSCs). For a thorough and detailed discussion of one S-4's experience in Somalia, see Michael (1994), p. 33.

CONCLUSIONS AND RECOMMENDATIONS

While deployments will always take their toll on equipment, some steps can be taken to mitigate the effects of this on MTW readiness while also addressing the unique requirements of POs. First, increased standardization of supply organizations and procedures in POs would reduce the uncertainty involved in obtaining replacements for unserviceable parts and equipment, and would facilitate implementation of the Army's Velocity Management initiatives.

Second, recognizing that certain equipment and materiel must be made available earlier in POs than in MTWs, and allowing staffs at the corps and division level to anticipate and plan these types of operations, could reduce the need for in-theater improvisation. This kind of planning can be conducted in the context of staff training exercises which involve developing TPFDLs and allocating constrained lift resources for a notional PO deployment.

Third, establishing PO-specific equipment sets to augment deploying units' TOE equipment would make crowd control, force protection and communications equipment, as well as nonlethal weapons and munitions, more readily available. These sets could be maintained at corps level. This recommendation may be difficult to accomplish in the face of resource constraints, but these capabilities are critical to PO success, and in some cases in short supply throughout the force, so should be worth the investment.

Fourth, the use of MTTs to train deploying soldiers to operate and maintain equipment with which they might not be familiar, such as crowd-control devices, radios, or nonlethal weapons, has proved useful in many deployments. Continuing, and perhaps expanding, this practice would help to minimize the effects of being unable to train routinely for the full range of possible missions.

Finally, for some multirotation POs, it may make sense to deploy a single equipment set with the initial units and then hand off to follow-on units. This has been done, for example, in Bosnia by the 362nd Combat Support Equipment Company from Fort Bragg, North

Carolina.[27] Until recently this approach would have gone against the Army practice of fighting with the equipment you train with, but the deployment to Kuwait of the 1st Cavalry Division in 1996 and routine NTC blue force rotations have demonstrated its potential. The costs of this approach include training soldiers on the use and maintenance of unfamiliar equipment, discomfort on the part of commanders who prefer to rely on the vehicles and weapons they know best, and degradation of stay-behind equipment as it sits in the motor pool.[28] But in some cases using prepositioned equipment sets or "single setting" through rotations could save the Army both time and money in terms of shipping and recovery, and could significantly decrease the redeployment time for a unit that has to move rapidly from a PO to an MTW.

_____

[27]Authors' interview with officers of the 20th Engineer Brigade and subordinate units, Fort Bragg, North Carolina, May 17, 1996.

[28]One option for providing consistent maintenance to such an equipment set might be to arrange for private contractor support for the duration of the operation.

Chapter Five

CONCLUSIONS

The Army is faced with a dilemma: it must maintain MTW readiness while frequently deploying to POs. Maintaining MTW readiness inevitably reduces PO capabilities; conducting POs inevitably reduces MTW readiness. Nonetheless, there are "win-win" steps that the Army can take to both improve PO performance and mitigate the effects of PO deployments on MTW readiness. Yet there are some intractable problems the Army faces that derive more from the military drawdown than from any specific kind of mission—though they are exacerbated by POs.

The win-win solutions include creating greater flexibility in the force structure, relying on intensive pre- and postdeployment training to help units transition from readiness for one mission to readiness for another, and continuing to implement Velocity Management throughout the Army. For POs, multirole and modular units are two ways to assist in tailoring the PO force while minimizing effects on MTW readiness. Modularity is intended to reduce the need for partial unit deployments for some kinds of support units, leaving stay-behind units intact and MTW-deployable; multirole units can increase the depth of available assets for certain requirements. Relying on non-Army capabilities (whether joint, coalition, interagency, or private contractors) also benefits both POs and MTWs. POs would gain from a more efficient and appropriate mix of capabilities, while key MTW assets (particularly certain support and special operations units) would be less likely to be stressed in multiple, prolonged PO deployments. Mobile training teams and intensive predeployment training exercises can help units tapped for PO deployments to prepare for their missions as efficiently as possible, without requiring them to spend significant amounts of time pursuing standardized PO training. Finally, continuing to implement Velocity Management will benefit both POs and MTWs by making supply and maintenance equally responsive to either contingency's requirements.

Such small changes may be more than sufficient. Thus far, the political mandates for peace operations have remained relatively limited; the military missions have been even more so. Nonetheless, there has been pressure in each operation for the Army to use its equipment and trained soldiers to support nation-building activities. Although this notion is disparaged as "mission creep" by some, there are those who consider greater Army involvement in civic action, humanitarian assistance, and civil-military relations to be key both to force protection and to creating the required environment for eventual withdrawal. Others simply see such an Army role as inevitable, since requirements for such efforts persist even under conditions where civilians cannot safely operate.

If the Army were directed to focus more on POs, force structure, training, and equipment would have to be modified to meet the greater requirements for policing, building infrastructure, and supporting a population.

In terms of force structure,

- Certain unit types would be needed in greater numbers (MPs, PSYOP and CA personnel, doctors, translators, lawyers, vertical engineers, and so forth) and depth.

- Training for commanders and staffs would have to reflect the increased emphasis on these kinds of activities.

- Equipment and personnel would have to be accessible for fine tailoring of units.

- Restructuring might also be necessary to prevent such force tailoring from stripping stay-behind units of key assets.

- Certain specialists, especially civil affairs and medical, would have to be made more readily available—either by creating more active slots or by designating particular reserve MOSs for easier and more frequent activation.

In terms of training and exercises,

- Light infantry divisions and their support forces, selected support forces from echelons above division and echelons above corps, and SOF (particularly PSYOP and CA), in addition to maintaining their combat skills, would have to go through regular combined training exercises to ensure effective command and coordination in POs.

- Military education and training would have to focus more systematically on such issues as interagency coordination, working with nongovernmental organizations, the requirements of negotiation and mediation, mine awareness, checkpoint and convoy skills, and so forth.

Finally, in terms of equipment, a commitment to peace operations would require, among other things,

- Better patrol vehicles (armored, wheeled, with gunports and mine protection).

- Sufficient demining equipment.

- Appropriate distribution and quantity of riot-control gear.

- Further development and procurement of nonlethal weapons.

- Sensitivity to the interoperability (across branches, services, government agencies, and nations) of communications and intelligence systems—both hardware and software.

Such steps would give the Army increased flexibility to respond appropriately to PO requirements. They could not be taken, however, without either shifting resources away from combat activities or increasing the overall Army budget.

Establishment of a Specialized PO Force Would Be Ill Advised

Some have suggested that the Army could establish a separate "peace operations force" (POF) that would train specifically for and deploy to PO missions, allowing combat and FSP units to maintain their focus on fighting and winning MTWs. This addresses many of the training concerns and also resolves some of the PO-related organizational and equipment issues. One thought is that the POF could be built around one of the existing armored cavalry regiments (ACRs), and the ACR headquarters could be expanded to include the command, control, communications, computer, intelligence, surveillance and reconnaissance (C^4ISR) assets needed to run fully synchronized logistics and small-unit combat operations.[1]

This approach has several deficiencies: (1) an unacceptably low level of combat training; (2) inadequate breadth of skills and resources to handle the full range of PO tasks; and (3) insufficient depth to take on a significant portion of the Army's PO burden. It should therefore be rejected.

An effective PO force must first be an effective combat force. The inability of UNPROFOR to gain the respect of the belligerents in Bosnia, the failure of UNOSOM I in Somalia, and the disintegration of the multinational force in Liberia illustrate the fate of any less-than-credible military force that interposes itself between warring parties. As Army leaders well know, combat effectiveness requires more than a part-time effort. But part-time combat training is precisely what the POF concept implies. Moreover, our investigation suggests that training for PO is not and need not be a major distractor from combat training.

POs span a broad range of missions from combat to humanitarian assistance. 10th Mountain Division infantry soldiers engaged in firefights in Mogadishu at the same time that quartermaster, medical, and transportation units were delivering relief supplies and services to starving Somalis. Combat units in Bosnia have had tense confrontations with the belligerent parties, but civil affairs teams and engineers have worked hand in hand with both Muslims and Serbs to restore municipal services and rebuild infrastructure. Such breadth suggests that the Army should find ways to draw more effectively on the strengths of *all* its units, rather than burdening a relatively small force with unrealistic expectations.

A historical review reveals that the number and scale of POs the Army has supported in recent years would quickly absorb all the assets of

---

[1]Authors' discussions with COL (Ret.) Don M. Snider and COL Daniel J. Kaufman, West Point, New York, November 19, 1996.

even a division-sized POF. Within just over a three-year period, Operation Restore Hope in Somalia required one mechanized battalion from the 24th Infantry Division and six light infantry battalions from the 10th Mountain Division, Operation Uphold Democracy in Haiti required the same force, and the follow-on force included an armored cavalry squadron and three light infantry battalions from the 25th Infantry Division. These operations were significantly smaller and shorter than the ongoing mission in Bosnia. Over the first 13 months, Operation Joint Endeavor employed more than ten armored or mechanized battalions under the command of two different divisions. The extent of CS/CSS and SOF forces demanded by these operations required the Army to draw on all its active component strength and to mobilize reserve and National Guard units.

Success in POs requires that participating units be just as combat-ready as for MTWs, and the diversity of PO activities demands task-organizations tailored to fit the specifics of the situation. Even if a unit could be organized and trained to be both combat-ready and expert in a wide range of PO tasks, its limited resources would quickly be overwhelmed by competing claims on its services. Rather than establishing a specialized POF, our research suggests that the Army can best address its dilemma by improving the flexibility, training, and equipment of its current forces.

Declining Resources Continue to Pose Challenges

Problems resulting from declining Army budgets complicate these policy issues. The cost-saving measure of maintaining units below strength will result in the need to cross-level for deployments, regardless of their nature: Operation Desert Storm required cross-leveling, just as operations in Haiti and Bosnia did. If resources were more plentiful, the expenditure of equipment in PO deployments would be less of an issue, since stockpiles could be increased to meet the demands of both MTWs and other contingencies. These are obviously resource, rather than PO, issues. Yet because PO deployments have been frequent and resources have remained constrained, the unintended outcome has been to meet PO needs at the expense of MTW readiness in some key areas.

In contrast, the collective combat training dilemma cannot be resolved with additional resources. Nor is it simply a question of priorities—deployments to POs will continue to interfere with collective combat training, both for deployed and stay-behind units. This may be the most intractable of the problems, though its effects on MTW readiness are ultimately situational. They could range from minor to serious, depending on the contingencies' proximity to each other in time and place, their natures, and so forth.

The Army's current approach is to treat each PO as an exception and to do little routine preparation for such contingencies. This is partly because the Army's size and budget in the past allowed it to absorb the costs of MOOTW. But a more fundamental determinant may be that assuming the risk of failure in a PO is much more generally acceptable than assuming the risk of failure in a major war. The result, however, is a second-order problem: PO deployments, rather than PO preparation, place undue stress on specific units, impede collective combat training, and decrease equipment availability and readiness for MTWs. The effects can probably be absorbed by the rest of the Army in the case of an MTW, with the exception of key low-density or frequently deployed units.[2] But attention must be paid to these issues, as well as to the longer-term issues of retention and enlistment (in both the AC and the RC, with particular attention to low-density MOSs), to ensure that capabilities remain balanced.

Clearly, given current constraints on resources and a continuing emphasis on MTWs, neither substantial changes to force structure, training, and equipment to improve PO performance, nor the current practice of responding to POs on an ad hoc basis, are ideal approaches. Fortunately, minor changes to force structure, training, and equipment will significantly improve PO performance without reducing combat readiness. Moreover, some of these changes also mitigate the effects of PO deployments on MTW readiness, producing "win-win" solutions for the Army.

---

[2]For example, Major General John Lemoyne, USAREUR's Deputy Chief of Staff for Operations and Plans (soon to be USAREUR Chief of Staff), recently told defense reporters that the command's modernization, training, and readiness remain on track, despite its year and a half of involvement in Bosnia. Caires (1997), p. 12. That is largely consistent with the findings of the General Accounting Office study on readiness, which, however, found that some civil affairs, PSYOP, signal, and transportation units were suffering degraded readiness because one or more of their subordinate elements had deployed to Bosnia. U.S. General Accounting Office (1996d), pp. 3–4.

PROJECT DATABASE

In attempting to condense and manage the data we collected from the literature search, case studies, and interviews, we created a qualitative database. It is organized by unit type, mission, citation, general and specific topic headings, and notes. This allowed us to search at a very detailed level. For example, we could look for all the information gathered for a specific unit type, or on a topic such as predeployment, or even by a single word, such as "transportation." We could run comparisons relatively easily as well: for example, we could search for all the notes we had gathered on low-density MOSs by case study, or for all our information on equipment shortfalls by unit type. The illustration on the following page is a representative slice of the database, to give the reader a more graphic sense of its organization and potential.[1]

---

[1]The database is not suitable for distribution, unfortunately. Since it was strictly for use by the project team, its Notes section contains shorthand explanations, placeholders, and inconsistent abbreviations and terminology that sufficed for the purposes of this project but make it problematic for broader use.

RAND MR921-A.1

Unit type	Mission	Cite	General	Specific	Notes
Engineer	IFOR	Interview, MAJ Barge	Force Structure	FS Requirements	Redhorse and Seabees 'saved the day' for bare base development, AF now doing a lot of construction for the Army. The Army should use its reserve construction units.
Engineer	Peace Ops	Interview, MAJ Brinkley/CPT Slaven	Force Structure	FS Requirements	Other armies, because their engineer forces are smaller, tend to have more generalized engineer units. US engineers may be overspecialized for peace ops.
Infantry	OAS	Interview, MAJ Brian Stapleton, 3rd Bn, 5th Cav, 1st Bde, 1st AD, by MAJ Daniel	Force Structure	FS Requirements	Engineer equip gets used very heavily, tends to break a lot. None of division mechanics were very familiar with that equip, but had civilian logistics assistance representative who knew eng equip and showed them how to fix it.

Had most organic mechanics with me, maintenance technician, Bn maintenance SGT, some PLL clerks who worked requisitions, etc. Was a scaled-down version of what we normally have in an Inf Bn. Had some support mechanics from 503rd Support Bn who did a lot of direct support maintenance. Not all mechanics belonged to 3-5 Cav, originally assumed they were all organizational mechanics. Found out, happily, that some did second echelon maintenance. |
| Infantry | OAS | Interview, SPC Chris Dobbins, Rifleman, D Co, 3rd Bn, 5th Cav | Force Structure | FS Requirements | Company CDRs first asked who wanted to go. Some people were short, were asked if they wanted to extend. Had to split the units up, took most of the good soldiers, left some behind to run the bn rear. |
| PSYOP | General | Interview, MAJ Craig Englehart Cowell, 9th BN, 5/16/96 | Force Structure | FS Requirements | Says we need more AC TAC BNs to support contingencies AND exercises

Need a lot of linguists for translation |
Engineer	General	Interview, 20th Eng Bde	Force Structure	LD MOS	Specialty MOSs are already shorthanded even before cross-leveling; topo surveyors and terrain analysts are two examples; also short on supply, mechanics.
CA	General	Interview, LTC Mike Rose, 5/15/96	Force Structure	LD MOS	The 96th CA BN's the only CA BN in the world with the global mission. Although the 422nd RC CA is dedicated to the 18th ABN worldwide, the 96th's NOT focused on any one unit in the Army; supports ANY unit, ANY CINC—only 5 co.s to watch the whole world
Engineer	General	Interview, 20th Eng Bde	Force Structure	LD MOS	Can't do collective training anymore because not enough people in specialized MOSs -- can't do field survey with four people, can't teach management skills

Specialty MOSs are already shorthanded even before cross-leveling: topo surveyors and terrain analysts are two examples; also short on supply, mechanics.

BIBLIOGRAPHY

Abizaid, John P., "Lessons for Peacekeepers," *Military Review*, March 1993.

Adelsberger, Bernard, "Combatting the Peacekeeping Stress," *Army Times*, May 27, 1996a.

———, "Griffith: 'Endeavor' Hampers Two-War Strategy," *Army Times*, May 6, 1996b.

Archambault, LTC Raoul, III, "Joint Operations in Haiti," *Army*, November 1995, pp. 22–29.

Atkinson, Rick, "Warriors Without a War: U.S. Peacekeepers in Bosnia Adjusting to New Tasks—Arbitration, Bluff, Restraint," *Washington Post*, April 14, 1996.

Barham, J. P., "Covering-Force Unit Is Ready, General Says," *European Stars and Stripes*, October 11, 1996.

Bender, Bryan, "Attack Helicopters Fill Reconnaissance Role in Bosnia," *Defense Daily*, October 17, 1996a.

———, "Nash: Bosnia Has Placed Army Firmly on Road to 21st Century," *Defense Daily*, October 16, 1996b.

"Bosnia Affecting Readiness But Army Still Able to Execute Two MRCs," *Inside the Army*, April 29, 1996.

Boyd, BG Morris J., "Peace Operations: A Capstone Doctrine," *Military Review*, May–June 1995.

Breen, Tom, "Bosnia Mission Improves Army Readiness, Joulwan Says," *Defense Daily*, October 16, 1996.

Brinkley, MAJ Phillip L., *Tactical Requirements for Peacekeeping Operations*, Fort Leavenworth, Kansas: School of Advanced Military Studies, U.S. Army Command and General Staff College, December 2, 1985.

Brown, Roger Allen, et al., *Assessing the Potential for Using Reserves in Operations Other Than War*, Santa Monica, California: RAND, MR-796-OSD, 1997.

Caires, Greg, "Army 'Better Off' Having Done Bosnia Mission, General Says," *Defense Daily*, July 2, 1997.

Caldwell, Jim, "In-Theater Soldier Training," *Army Trainer Magazine* (on-line), November 30, 1995.

"Commander of Haiti Operation Says Peacekeeping Dulls Warfighting Edge," *Inside the Army*, May 13, 1996.

Davis, Lois M., Susan D. Hosek, Michael G. Tate, et al., *Army Medical Support for Peace Operations and Humanitarian Assistance*, Santa Monica, California: RAND, MR-773-A, 1997.

Department of the Army, *Training the Force: Battle Focused Training: Battalion and Company Soldier, Leaders & Units*, Field Manual 25-101, Washington, D.C.: Headquarters, Department of the Army, September 1990.

———, *Operations*, FM 100-5, Washington, D.C.: Headquarters, Department of the Army, June 14, 1993a.

———, *U.S. Army Forces Somalia: 10th Mountain Division (LI) After Action Report*, Department of the Army, June 2, 1993b.

———, *Peace Operations*, FM 100-23, Washington, D.C.: Headquarters, Department of the Army, December 30, 1994.

———, TRADOC Analysis Center, *Reconstitution of Army Combat Service Support Units Engaged in Operations Other Than War: Or, Do We Have Enough To Fulfill the National Military Strategy?* Fort Lee, Virginia: Department of the Army TRADOC Analysis Center, Technical Report TRAC-TR-0995, February 1996.

Department of the Army, Department of the Air Force, *Military Operations in Low Intensity Conflict*, FM 100-20/AFP 3-20, Washington, D.C.: Headquarters, Departments of the Army and the Air Force, December 5, 1990.

Dworken, Jonathan T., "Rules of Engagement: Lessons from Restore Hope," *Military Review*, September 1994.

Fisher, Ian, "For G.I.'s, More Time Away From Home Fires," *The New York Times*, December 24, 1996.

Goodman, Glenn W., Jr., "Civil Savvy in Special Ops: Skills of Army Civil Affairs Personnel in Demand," *Armed Forces Journal International*, October 1994.

Graham, Bradley, "Revamped National Guard: No Cuts But More Support Jobs," *Washington Post*, November 7, 1996.

Haskell, Bob, "A Total Force Effort for Peace," *Army*, May 1996, pp. 35–36.

"Impacts of Operations Other Than War (OOTW) on Unit Readiness," *Memorandum for Commander, XVIII Airborne Corps*, February 1995.

Kinzer, S., "Dutch Conscience Stung by Troops' Bosnia Failure," *The New York Times*, October 8, 1995, pp. 1, 12.

Kirkland, Faris R., Ronald R. Halverson, and Paul D. Bliese, "Stress and Psychological Readiness in Post–Cold War Operations," *Parameters*, Vol. 26, No. 2, Summer 1996.

Komarow, Steve, "Shalikashvili Reshaping Strategy," *USA Today*, May 13, 1996, pp. 1, 13.

Litz, Brett, "The Psychological Demands of Peacekeeping for Military Personnel," *Clinical Quarterly*, Vol. 6, Issue 1, Winter 1996.

MacFarland, Margo, "Rethinking 'Tooth-to-Tail': New Missions Demand New Approaches," *Armed Forces Journal International*, September 1, 1994.

MacKinlay, J.C.G., "Multinational Peacekeeping Forces," *RUSI Journal*, December 1983.

"Mailcall: Is It Possible to Be Mobilized Twice?" *Army Reserve Magazine*, Fall 1996.

Michael, Stephen, "CSS Operations in Somalia," *Infantry Magazine*, July–August 1994.

Miller, Laura. "Do Soldiers Hate Peacekeeping? The Case of Preventive Diplomacy Operations in Macedonia," *Armed Forces and Society*, Vol. 23, No. 3, Spring 1997, pp. 415–450.

Nagy, COL William J., "CSS Force Multiplier," *Army Reserve Magazine*, Fall 1996.

"New Mission in Bosnia has Same Dangers," *USA Today*, October 22, 1996.

Niblack, Preston, "The United Nations Mission in Haiti: Trip Report," unpublished trip report, RAND, 1995.

O'Connor, Mike, "Does Keeping the Peace Spoil G.I.'s for War?" *The New York Times*, December 13, 1996, p. 1.

Palumbo, Sue, "Unit Moving On Out—and Heading Home," *European Stars and Stripes*, August 16, 1996.

Peace Enforcement Operations, Lessons Learned Newsletter #1-95, MOUNTAIN EAGLE 95, USAREUR.

Peacekeeping, Canadian Forces Command and Staff College, Command and Staff Course 4, 1978.

Peters, Katherine McIntire, "Price of Peace for Reservists: A Stable Life Back at Home," *Army Times*, January 1, 1996, p. 28.

Pexton, Patrick, "Study: Troops Half-hearted on Peacekeeping," *Army Times*, No. 18, November 27, 1995.

Pilgrim, LTC Calvin, "Logisticians Execute Peace Operations," *Army Logistician*, January–February 1996.

Reese, Robert J., *Operational Considerations for the Employment of a Light Infantry Division in a Contingency Scenario*, Fort Leavenworth, Kansas: School of Advanced Military Studies, U.S. Army Command and General Staff College, May 21, 1987.

Ritchie, E. Cameron, David C. Ruck, and Milton, W. Anderson, "The 528th Combat Stress Control Unit in Somalia in Support of Operation Restore Hope," *Military Medicine*, Vol. 159, May 1994.

Roos, John G., "Early Lessons from Haiti: A Non-War Can Devour the 'Force of Choice,'" *Armed Forces Journal International*, November 1994.

Schroeder, LTG Daniel, "Lessons of Rwanda: Joint Warfighting Doctrine Works in Operations Other Than War," *Armed Forces Journal International*, December 1994, pp. 31–33.

Science Applications International, *Peace Operations Training and Education in the U.S. Armed Forces*, McLean, Virginia: Science Applications International Corporation, November 18, 1994.

Segal, D. R., and M. W. Segal (eds.), *Peacekeepers and Their Wives*, Westport, Connecticut: Greenwood Press, 1993.

Segal, David R., Jesse J. Harris, Joseph M. Rothberg, and David H. Marlowe, "Paratroopers as Peacekeepers," *Armed Forces and Society*, Vol. 10, No. 4, Summer 1984.

Segal, David R., Theodore P. Furukawa, and Jerry C. Lindh, "Light Infantry as Peacekeepers in the Sinai," *Armed Forces and Society*, Vol. 16, No. 3, Spring 1990, pp. 385–403.

Seventh Army Training Command, "CMTC Lessons Learned: Contingency Operations," Training Note No. 9, Grafenwoehr, Germany, July 1993.

Shelton, LTG Henry, "Impact of Operations Other Than War on Training and Readiness as Related to Recovery Time," unpublished memo to GEN Dennis J. Reimer, February 22, 1995.

Shelton Interview Transcript, AHAS, Ft. Leavenworth, February 13, 1995.

Sortor, Ronald E., *Army Forces for Operations Other Than War*, Santa Monica, California: RAND, MR-852-A, 1997.

Stanton, MAJ Martin N., "Task Force 2-87: Lessons from Restore Hope," *Military Review*, September 1994.

Stennett, Rick, and Jim Walley, "Operations Other Than War, Volume IV, Peace Operations," *Center for Army Lessons Learned Newsletter* No. 93-8, December 1993, Chapter V.

Story, Ann E., and Aryea Gottlieb, "Beyond the Range of Military Operations," *Joint Forces Quarterly,* Autumn 1995, pp. 99–104.

Sullivan, Jerry E., "The Army Reserve in Bosnia," *Army,* April 1996.

Summers, Harry, "Military Urban Myths," *Washington Times,* October 24, 1996, p. 16.

Task Force Able Sentry Training, Annex T to V Corps OPORD 95-55.

"Training Program for Staffs," Operations Group Delta, Battle Command Training Program, Fort Leavenworth, Kansas, training package introductory pamphlet.

U.S. Army Center for Army Lessons Learned, *U.S. Army Operations in Support of UNOSOM II, Lessons Learned Report,* Fort Leavenworth, Kansas: Center for Army Lessons Learned, 4 May 1993–31 March 1994.

———, *Operation Uphold Democracy: Initial Impressions,* Fort Leavenworth, Kansas: Center for Army Lessons Learned, December 1994.

———, *Peace Operations Training Vignettes with Possible Solutions,* Newsletter No. 95-2, Fort Leavenworth, Kansas: Center for Army Lessons Learned, March 1995a.

———, *The Effects of Peace Operations on Unit Readiness, Interim Report,* Fort Leavenworth, Kansas: Center for Army Lessons Learned, June 1995b.

———, *The Effects of Peace Operation on Unit Readiness,* Fort Leavenworth, Kansas: Center for Army Lessons Learned, February 1996a.

———, *Operation Joint Endeavor: Initial Impressions Report,* Fort Leavenworth, Kansas: Center for Army Lessons Learned, July 1996b.

U.S. Army Concepts Analysis Agency, *Force Employment Study*, Bethesda, Maryland: U.S. Army Concepts Analysis Agency, February 1991.

U.S. Army Pacific (USARPAC) METL, document from USARPAC internet homepage, Headquarters, USARPAC, Hawaii, undated (May 1996).

U.S. Department of Defense, *Report to the House Appropriations Committee: Support to Operations Other Than War and Its Impact on TEMPO Rates*, Washington, D.C.: Department of Defense, June 1996.

U.S. Department of Defense, *Memorandum for Correspondents*, 002-M, URL http://www.dtic.dla.mil/defenselink/news/Jan97/m011097_m002-97.html, January 10, 1997.

U.S. Forces Command (FORSCOM) Proposed METL, III Corps briefing chart, Headquarters, U.S. FORSCOM, Fort McPherson, Georgia, undated (May 1996).

United States Stability Operations in the Dominican Republic 28 April 1965–30 May 1965, Part I, Volume IV, Chapter 17: Doctrine and Force Organization, Headquarters, U.S. Forces Dominican Republic, August 31, 1965.

U.S. General Accounting Office, *Peace Operations: Withdrawal of U.S. Troops from Somalia*, Washington, D.C.: General Accounting Office, GAO/NSIAD-94-175, June 1994.

——, *Peace Operations: Effect of Training, Equipment, and Other Factors on Unit Capability*, Washington, D.C.: General Accounting Office, GAO/NSIAD-96-14, October 1995a.

——, *Peace Operations: Heavy Use of Key Capabilities May Affect Response to Regional Conflicts*, Washington, D.C.: General Accounting Office, GAO/NSIAD-95-51, March 1995b.

——, *Military Readiness: A Clear Policy Is Needed to Guide Management of Frequently Deployed Units*, Washington, D.C.: General Accounting Office, GAO/NSIAD-96-105, April 1996a.

————, *Contingency Operations: Defense Cost and Funding Issues,* Washington, D.C.: General Accounting Office, GAO/NSIAD-96-121BR, March 1996b.

————, *Military Readiness: Data and Trends for January 1990 to March 1995,* Washington, D.C.: General Accounting Office, GAO/NSAID-96-111BR, March 1996c.

————, *Military Readiness: Data and Trends for April 1995 to March 1996,* Washington, D.C.: General Accounting Office, GAO/NSAID-96-194, August 1996d.

————, *Peace Operations: Reservists Have Volunteered When Needed,* Washington, D.C.: General Accounting Office, GAO/NSIAD-96-75, April 1996e.

————, *Force Structure: Army Support Forces Can Meet Two-Conflict Strategy With Some Risks,* Washington, D.C.: General Accounting Office, GAO/NSIAD-97-66, February 1997a.

————, *Military Readiness: Improvements Still Needed in Assessing Military Readiness,* Testimony before the Subcommittee on Military Readiness, Committee on National Security, House of Representatives, Washington, D.C.: General Accounting Office, GAO/T-NSIAD-97-107, March 1997b.

"Wargames Show Tension Between Overseas Engagement, OPTEMPO Limits," *Inside the Pentagon,* June 16, 1997, p. 1.

Weisaeth, L., P. Aarhaug, L. Mehlum, and S. Larsen, *The UNIFIL Study (1991–1992), Report I: Results and Recommendations,* Norway: Headquarters Defence Command Norway, the Joint Medical Service, 1993.

Weisaeth, Lars, Lars Mehlum, and Mauritz S. Mortensen, "Peacekeeper Stress: New And Different?" *Clinical Quarterly,* Vol. 6, Issue 1, Winter 1996.

West, Togo D., Jr., and GEN Dennis J. Reimer, *A Statement on the Posture of the United States Army, Fiscal Year 1998,* http://www.army.mil/aps/chapter2.htm, February 1997.

Willis, G. E., "From the Top," *Army Times,* July 15, 1996a, pp. 12–14.

————, "On the Road Again," *Army Times*, July 1, 1996b, pp. 12–14.

Williams, Steve, "Senior-Level Leadership Responsibilities in Operations Other Than War," Lesson 5, Course C520, *Operations Other Than War*, U.S. Army Command and General Staff College, January 1996.

Wiseman, Henry, *Peacekeeping: Appraisals and Proposals*, New York: Pergamon Press, 1983.

Wong, Leonard, Paul D. Bliese, and Ronald R. Halverson, "Multiple Deployments: Do They Make a Difference?" paper presented at the Biannual Conference of the Inter-University Seminar on Armed Forces and Society, Baltimore, October 1995.

Woodberry, John, "Soldiering for Peace: U.S. Combat Training for Non-Combat Operations in Bosnia," *Jane's International Defense Review*, April 1996.

Zvijac, David J., and Katherine A. W. McGrady, *Operation Restore Hope: Summary Report*, Alexandria, Virginia: Center for Naval Analyses, CRM 93-152, March 1, 1994.

Mission-Essential Task List (METL) Documents

I Corps METL, "Enclosure 1, I Corps Mission Essential Task List, to FY 96–97 Training Guidance," Headquarters, I Corps, Fort Lewis, Washington, undated (May 1996).

III Corps METL, briefing chart, Headquarters, III Corps, Fort Hood, Texas, undated (May 1996).

V Corps METL, 3rd Infantry Division [since redesignated 1st Infantry Division] briefing chart, Headquarters, V Corps, Heidelberg, Germany, undated (November 1995).

XVIII Airborne Corps METL, memorandum subject: "Fiscal Year (FY) 1997–1998 Command Training Guidance," Headquarters, XVIII Abn Corps, Fort Bragg, North Carolina, undated (1995).

1st Armored Division METL, briefing chart, Headquarters, 1st Armored Division, Bad Kreuznach, Germany, undated (November 1995).

1st Brigade, 25th Infantry Division METL, briefing chart titled "Commander's METL Assessment," Headquarters, 1st Brigade, 25th Infantry Division, Fort Lewis, Washington, undated (May 1996).

1st Cavalry Division METL, memorandum subject: "Division Mission Statement and Mission Essential Task List (METL)," Headquarters, 1st Cavalry Division, Fort Hood, Texas, April 1, 1996.

1st Infantry Division METL, memorandum subject: "3d Infantry Division [since redesignated 1st Infantry Division] Mission Essential Task List," Headquarters, 1st Infantry Division, Wurzburg, Germany, November 7, 1995.

2nd Infantry Division METL, memorandum subject: "Third Quarter FY96 Training Guidance," Headquarters, 2nd Infantry Division, Korea, January 31, 1996.

"3d POB (A) PSYOP PERSONNEL CAPABILITY," unpublished document, May 1996.

3rd Brigade, 2nd Infantry Division METL, briefing chart titled "3BCT Mission Statement," Headquarters, 3rd Brigade Combat Team, Fort Lewis, Washington, undated (May 1996).

3rd Personnel Group METL, memorandum subject: "3d Personnel Group Mission Essential Task List (METL)," Headquarters, 3rd Personnel Group, Fort Hood, Texas, April 26, 1996.

3rd Psychological Operations Battalion (DIS) (Airborne) METL, memorandum subject: "3d POB(A) Mission Statement and METL," 3rd Psychological Operations Battalion (DIS) (AIRBORNE), Fort Bragg, North Carolina, October 12, 1995.

3rd Signal Brigade METL, memorandum subject: "Revision of 3d Signal Brigade Mission Essential Task List," Headquarters, 3rd Signal Brigade, Fort Hood, Texas, May 8, 1996.

4th Infantry Division METL, memorandum subject: "Revised Mission Essential Task List (METL)," with attached briefing chart, Headquarters, 4th Infantry Division, Fort Hood, Texas, April 17, 1996.

6th Psychological Operations Battalion (Airborne) METL, memorandum subject: "6th Psychological Operations Battalion (Airborne) Mission Statement and METL," 6th Psychological Operations Battalion (Airborne), Fort Bragg, North Carolina, November 20, 1995.

Eighth U.S. Army (EUSA) METL, 175th Finance Command briefing chart, Headquarters, EUSA, Korea, undated (May 1996).

9th Psychological Operations Battalion METL, briefing chart titled "Commander's METL Assessment," Headquarters, 9th PSYOP Bn, Fort Bragg, North Carolina, undated (May 1996).

13th Corps Support Command METL, memorandum subject: "Proposed COSCOM METL," with attached briefing chart, Headquarters, 13th COSCOM, Fort Hood, Texas, April 22, 1996.

44th Medical Brigade METL, document from 44th Medical Brigade internet homepage, Headquarters, 44th Medical Brigade, Fort Bragg, North Carolina, undated (May 1996).

62nd Medical Group METL, briefing chart titled "Quarterly Training Brief," Headquarters, 62nd Medical Group, Fort Lewis, Washington, May 15, 1996.

201st Military Intelligence Brigade METL, briefing chart, Headquarters, 201st MI Bde, Fort Lewis, Washington, undated (May 1996).

504th Military Police Battalion METL, briefing chart, Headquarters, 504th Military Police Battalion, Fort Lewis, Washington, undated (May 1996).

U.S. Army Combat Aviation Training Brigade METL, memorandum subject: "MSC METL Development," with attached briefing charts, Headquarters, USA CATB, Fort Hood, Texas, May 15, 1996.

U.S. Army Europe (USAREUR) METL, document from USAREUR internet home page, Headquarters, USAREUR, Heidelberg, Germany, undated (May 1996).